Badger GCSE Science
How Science Works

Biology

Andrew Grevatt
Dr. Deborah Shah-Smith

You may copy this book freely for use in your school.
The pages in this book are copyright, but copies may be made without fees or prior permission provided that these copies are used only by the institution which purchased the book. For copying in any other circumstances, prior written consent must be obtained from the publisher.

Badger Publishing

INTRODUCTION

This is the first of a series of three books of 'How Science Works' activities for GCSE science. These have been developed by teachers to give them a range of resources to use in teaching science through the How Science Works principles. These activities work well as an 'add on' to any of the KS4 schemes of work.

How Science Works is now an essential part of the Science National Curriculum. For many science teachers, this has meant getting to grips with a new approach to science teaching. These activities have been developed to encourage structured discussion, improve knowledge and understanding of How Science Works and support learners to consider a range of viewpoints and make an informed decision.

The new KS4 Science National Curriculum sets out the parameters for How Science Works and the specifications for each of the GCSE examination boards have interpreted these in a variety of ways. We have developed these activities to be suitable for use in all schools, whatever exam board they use.

We have found that learners do not necessarily have the background knowledge about scientific issues such as genetically modified food. Without this background information, it is hard for them to discuss the issue and form an opinion. These How Science Works activities have been developed to help boost background knowledge of scientific issues, consider evidence from a range of contexts and support learners to make a decision.

HOW TO USE THESE ACTIVITIES

The general approach to using these activities is to introduce the learners to the task, then allow them to discuss the task in pairs or groups of four. During this time, the teacher should circulate amongst the groups to encourage discussion. Once the learners have had time to discuss their ideas and make decisions, the teacher can lead a discussion to draw out the key points.

More specifically, the activities take a variety of forms that require slightly different management. **Data analysis, graph drawing and predicting activities** follow the general approach but note that the discussion of the activities should focus on the *process* of analysing, data presentation and prediction rather than just what the answer is.

Card sorts take a variety of forms, but the general approach allows learners to physically sort statements into groups. Discussion should focus on the nature of evidence or facts and opinions in the context of the content of the activity.

Timelines require more time and can be extended by allowing learners to add their own images and additional events on the timeline. They are designed so that learners can identify how ideas change and relate images to the text. Follow-up discussion should focus on how and why the ideas have changed and perhaps imagining what may happen in the future.

A NOTE ABOUT TIMING

We have identified the approximate time taken for the activities based on an average ability class. More able groups may need less time on the task but more time on the discussion and lower ability learners may need support (particularly literacy) through the activities themselves. Most tasks have some differentiation suggestions.

AUTHORS

Andrew Grevatt is an experienced Advanced Skills Teacher who is currently an associate tutor at the University of Sussex, where he is researching for a professional doctorate in education. He would like to dedicate this publication to his newest nephew, Toby William.

Dr. Deborah Shah-Smith is an experienced science teacher. She has a keen interest in developing resources using practice-based evidence. She would like to dedicate this publication to her husband, Paul, and daughter, Zaveri.

ACKNOWLEDGEMENTS

We would like to thank our colleagues who have helped us to develop these tasks. These include Ben Riley of Oriel High School, West Sussex, Ross Palmer of Cardinal Newman School, Brighton and Hove.

Contents and Curriculum Links

INTERDEPENDENCE AND ADAPTATION	NC HSW		ACTIVITY TYPE
1. Population Predictions	1abc	Data, evidence, theories and explanations	Graph analysis and predictions
2. Predicting Fish Populations	1abcd	Data, evidence, theories and explanations	Graph analysis and predictions
3. Organic Prices	1ab	Data, evidence, theories and explanations	Data analysis
4. Where Limpets Live	1ab	Data, evidence, theories and explanations	Graph analysis
5. Food Chain Calculations	2abd	Practical and enquiry skills	Planning using secondary resources
6. Intensive Farming: Social, Economic and Environmental Issues	4b	Applications and implications of science	Card sort and discussion
7. Organic Farming: Benefits, Drawbacks and Risks	4a	Applications and implications of science	Card sort and discussion

VARIATION AND EVOLUTION			
8. How We Have Grown!	3abc	Communication skills	Graph drawing and analysis
9. Ivy Leaves	1bc	Data, evidence, theories and explanations	Predicting
10. Dinosaur Extinction	1abcd	Data, evidence, theories and explanations	Evaluating evidence
11. Fossil Evidence	1abcd	Data, evidence, theories and explanations	Evidence card sort
12. Evolution of the Horse	3ab	Communication skills	Evidence card sort
13. How to Clone	3ac	Communication skills	Card sequencing
14. Designer Babies	4abc	Applications and implications of science	Discussion and decision making
15. Evolution Timeline	4c	Applications and implications of science	Card sequencing

GENES AND GENETICS			
16. Testing Antibiotics	3abc	Communication skills	Data analysis
17. Genetic Modification of Crops: Fact or Opinion?	4c	Applications and implications of science	Evidence card sort
18. Industrial Fermenters	2ab	Practical and enquiry skills	Matching card sort
19. How Gene Splicing Works	2ab	Practical and enquiry skills	Card sequencing
20. DNA Fingerprint Analysis	3ab	Communication skills	Data analysis
21. Genetics Timeline	4c	Applications and implications of science	Card sequencing
22. Growing GM Cotton in India: Social, Economic and Environmental Issues	4b	Applications and implications of science	Card sort and discussion
23. Genetically Modified (GM) Crops: Benefits, Drawbacks and Risks	4a	Applications and implications of science	Card sort and discussion

ELECTRICAL AND CHEMICAL SIGNALS	NC HSW		ACTIVITY TYPE
24. Measuring Reaction Times	2d	Practical and enquiry skills	Data analysis
25. Driver Response Times	2bd	Practical and enquiry skills	Data analysis
26. Tired Reactions	2ab	Practical and enquiry skills	Planning an investigation
27. Improving Reaction Times	3a	Communication skills	Using scientific knowledge
28. Blood Sugar Graphs	3a	Communication skills	Graph interpretation
29. Female Hormonal Graph Analysis	3abc	Communication skills	Graph interpretation
30. Diabetes Timeline	4c	Applications and implications of science	Card sort
31. The Contraceptive Pill: Social, Economic and Environmental Issues	4b	Applications and implications of science	Discussion and card sort
32. In Vitro Fertilisation (IVF): Benefits, Drawbacks and Risks	4a	Applications and implications of science	Discussion and card sort

HEALTH, MEDICINE AND DRUGS			
33. Cannabis and Tobacco – The Evidence	1abcd	Data, evidence, theories and explanations	Discussion and card sort
34. Drug Trials	1abc	Data, evidence, theories and explanations	Card sequencing
35. Smoking Statistics	3abc	Communication skills	Graph data analysis
36. Antibiotic Resistance	3abc	Communication skills	Graph drawing and analysis
37. Treating TB	3abc	Communication skills	Graph data analysis
38. Weights and Measures in Medicine	3bc	Communication skills	Units card sort
39. TB Timeline	4c	Applications and implications of science	Card sequence
40. Childhood Vaccinations: Social, Economic and Ethical Issues	4b	Applications and implications of science	Discussion and card sort
41. Stem Cell Therapy: Benefits, Drawbacks and Risks	4a	Applications and implications of science	Discussion and card sort

ENVIRONMENT			
42. Sampling: Quadrat, Transect or Satellite Imaging?	1a	Data, evidence, theories and explanations	Planning
43. Analysing Global Warming	1c	Data, evidence, theories and explanations	Data analysis and application
44. Quadrat Data Comparison	1a	Data, evidence, theories and explanations	Data analysis
45. Lichens and Air Pollution	3abc	Communication skills	Data analysis and application
46. Fertiliser Choices	1abc	Data, evidence, theories and explanations	Data analysis and application
47. Malaria and DDT	3abc	Communication skills	Data analysis
48. DDT: How Ideas Changed	4c	Applications and implications of science	Card sequence
49. Global Warming: Fact or Opinion?	4c	Applications and implications of science	Discussion and card sort
50. Pesticides: Benefits, Drawbacks and Risks	4a	Applications and implications of science	Discussion and card sort

INTERDEPENDENCE & ADAPTATION
POPULATION PREDICTIONS

1

KS4 NATIONAL CURRICULUM HSW LINK

1. Data, evidence, theories and explanations
   a. how interpretation of data, using creative thought, provides evidence to test ideas and develop theories
   b. how explanations of many phenomena can be developed using scientific theories, models and ideas
   c. that there are some questions that science cannot currently answer, and some that science cannot address

RESOURCES:
Task Sheet 1, enough for one per learner.

TIME:
10 minutes for the activity, 10 minutes for class discussion.

NOTES

- This activity is most suitable for use as a Starter, Main activity or Plenary.
- Learners with literacy issues may need to have the text read to them.
- Break down the task into two parts: comparison of the two graphs and then a discussion of Cohen's ideas.

SUGGESTED ANSWERS

A. What are the natural barriers on Earth to human population size? *Amount of food, clean water, space to grow food, actual space to live, diseases (the more densely packed we are, the more chance of disease spreading).*

B. What are the uncertainties that make it so difficult to predict the human carrying capacity? *Unpredictable climate. Humans are living longer and longer. We do not know what medical advances will be made in the future. We do not know the technological advances that may be made in buildings and houses (multi-storey buildings, sleeping pods, etc). There may be more efficient farming practices.*

C. What do you think the world would be like if humans reached their carrying capacity? *Possibly very crowded, lots of pollution and waste, high demands on medical services, reduced space to grow crops and rear animals for food, more chance of (new) diseases spreading.*

EXTENSION SUGGESTION

Consider what life would be like in Britain if the population doubled.

INTERDEPENDENCE & ADAPTATION
POPULATION PREDICTIONS

1

If we put a nutrient broth into a beaker and then add one bacterium, that bacterium will divide and the new bacteria will divide, using the nutrients in the broth. Will the bacteria be able to keep dividing forever? No. The population will reach a point where the amount of nutrients will run out and they will run out of space in the beaker, so no new bacteria can be produced. The point at which the bacteria reach their maximum population in the beaker is called the **carrying capacity** (Figure 1).

Figure 1 Growth curve of bacteria

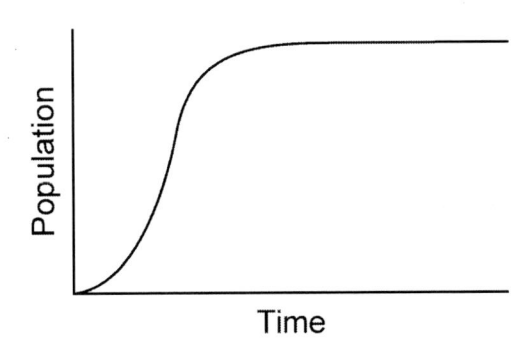

Figure 2 Population growth curve for humans

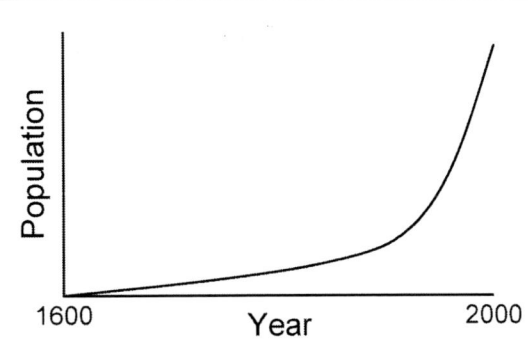

TASK

1. Compare the human population curve (Figure 2) with the bacteria population curve (Figure 1). Do you think that there is a human carrying capacity on the Earth? Read the views of this scientist and discuss them.

> Joel Cohen wrote a book, published in 1995, called *How many people can the Earth support?* He argues that:
> - Scientists find it very difficult to estimate the human carrying capacity.
> - Most estimates are of between 10-20 million people.
> - We need to consider the well-being of humans at the carrying capacity.
> - The solution to maximizing the human carrying capacity and a good level of well-being involves:
>   – Making the pie bigger: the more humans there are, the more creativity there will be to overcome the natural barriers to population size.
>   – Putting more forks on the table: reduce people's expectations, change to a vegetarian diet and improve family planning.
>   – Teach better manners: governments must improve the well-being of everyone.

Discuss
A. What are the natural barriers on Earth to human population size?
B. What are the uncertainties that make it so difficult to predict the human carrying capacity?
C. What do you think the world would be like if humans reached their carrying capacity?

INTERDEPENDENCE & ADAPTATION
PREDICTING FISH POPULATIONS

2

KS4 NATIONAL CURRICULUM HSW LINK

1. *Data, evidence, theories and explanations*
   a. how scientific data can be collected and analysed
   b. how interpretation of data, using creative thought, provides evidence to test ideas and develop theories
   c. how explanations of many phenomena can be developed using scientific theories, models and ideas
   d. that there are some questions that science cannot currently answer, and some that science cannot address

RESOURCES:
Task Sheet 2, enough for one per learner.

TIME:
10 minutes for the activity, 10 minutes for class discussion.

NOTES

- This activity is suitable for use as an extended Starter or Plenary, or Main or Homework activity.
- Learners must be aware of these key words before attempting the task: edible, population.

SUGGESTED ANSWERS

Although the main graph is based on published data, the three graphs to interpret are fictional. However, the purpose of the exercise is purely interpretation based within the context of populations. Note the x-axis scales change on Graphs 2 and 3.

A. Estimated population of edible fish = Graph 1 (inverse of Figure 1)
B. Estimated catch (millions of tonnes) = Graph 3 (assuming fish populations continue to decline, they will be harder to catch)
C. Average cost per tonne of fish caught (US$) = Graph 2 (as fish become more scarce, they will become more expensive)

EXTENSION SUGGESTION

List some suggestions of how to make sea fishing sustainable.

INTERDEPENDENCE & ADAPTATION
PREDICTING FISH POPULATIONS

2

Fishing boats catch fish from the sea for people to eat. This happens around the world. Look at Figure 1. It shows the estimate of the number of fish caught each year over 50 years.

Figure 1: Estimated global fish catches 1950–1999

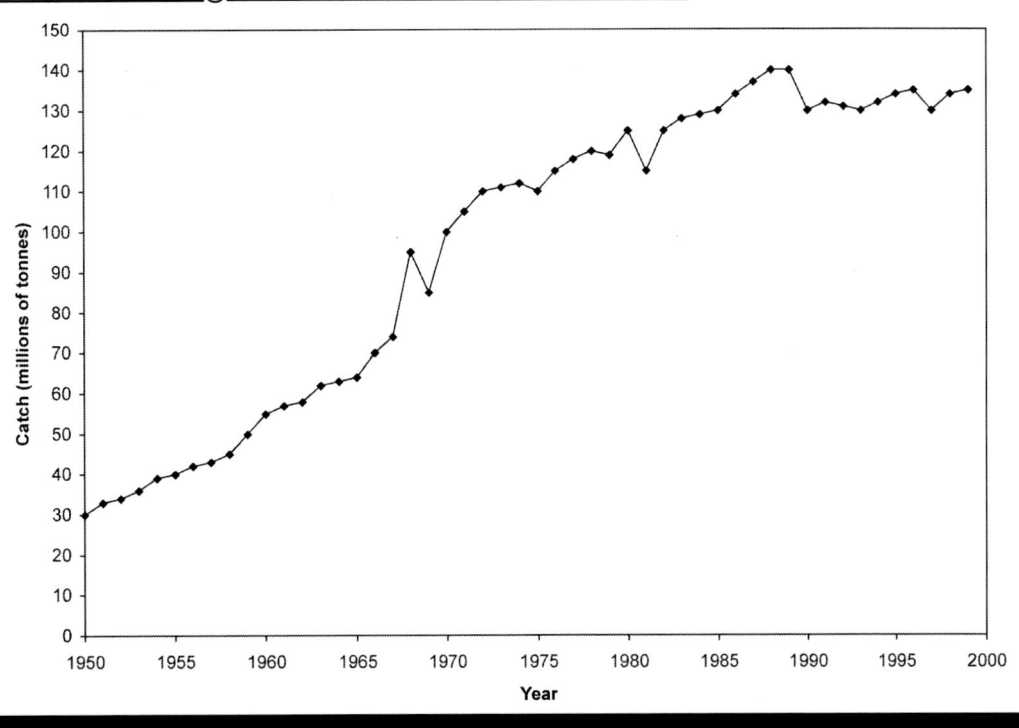

TASK

What is the general trend of the number of fish caught over the 50 years (Figure 1)?

Look at the graphs below and the *y*-axis labels (A, B and C).
Match the *y*-axis labels to the graphs. Give reasons for your choice.

Y-axis labels
A. Estimated population of edible fish.
B. Estimated catch (millions of tonnes).
C. Average cost per tonne of fish caught (US$).

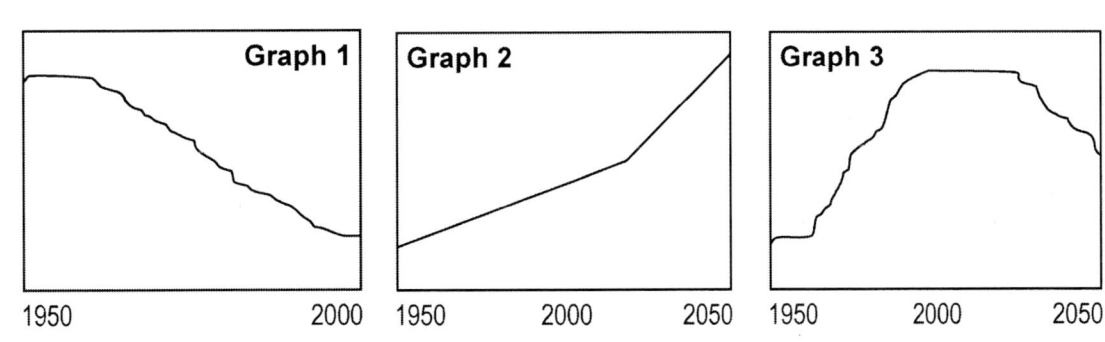

Graph 1 | Graph 2 | Graph 3

KS4 Biology How Science Works: Copymaster
Biology HSW © Badger Publishing Ltd

INTERDEPENDENCE & ADAPTATION
ORGANIC PRICES

3

KS4 NATIONAL CURRICULUM HSW LINK

1. Data, evidence, theories and explanations
   a. how interpretation of data, using creative thought, provides evidence to test ideas and develop theories
   b. how explanations of many phenomena can be developed using scientific theories, models and ideas

RESOURCES:
Task Sheet 3, enough for one per learner or one between two learners.

TIME:
10 minutes for the activity, 5-10 minutes for class discussion.

NOTES

- This activity is suitable for use as a Starter, Main activity or Plenary.
- Learners must be aware of these key words/concepts before attempting the task: intensive farming, organic farming.

SUGGESTED ANSWERS

A. Why is organic food more expensive than non-organic food?
   More space needed to rear the animals. No chemical pesticides or herbicides used, so more loss of crops to pests and disease (reduced yield), more hands-on work required for weeding, more time required. Animal breeds are less efficient.

B. Why is meat so much more expensive than bread?
   Points related to energy loss between each stage of the food chain. Shorter food chain for plants compared to animals, so less time and energy. Plants grow faster than 'meat'.

C. Why do some people prefer to eat organic food?
   Believe it is healthier as it has no pesticides in it that could be harmful to health. Animals are reared in a more humane way – more space, variety of feed, able to behave naturally.

EXTENSION SUGGESTION

Consider if organic food will ever come down in price. Explain your answers.

INTERDEPENDENCE & ADAPTATION
ORGANIC PRICES

3

TASK

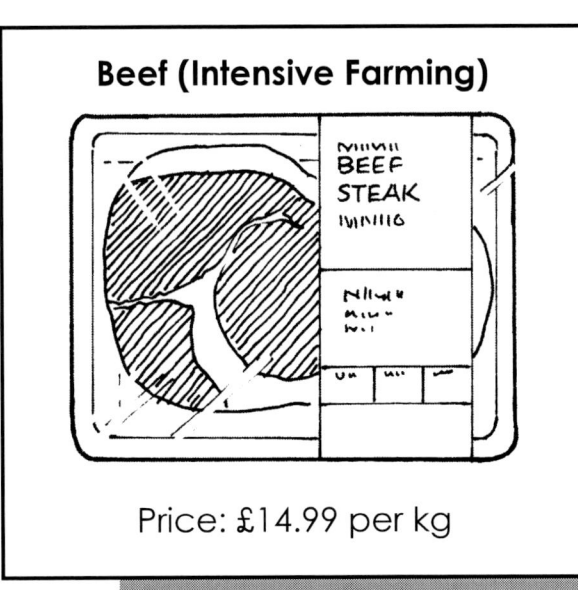

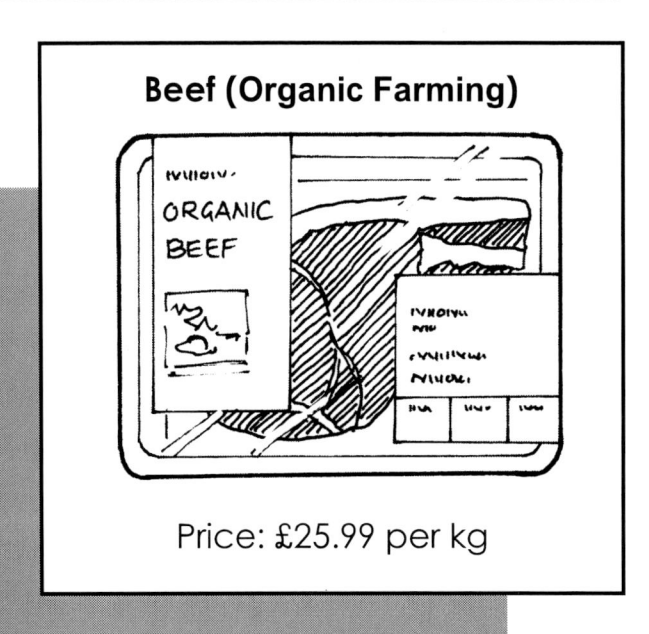

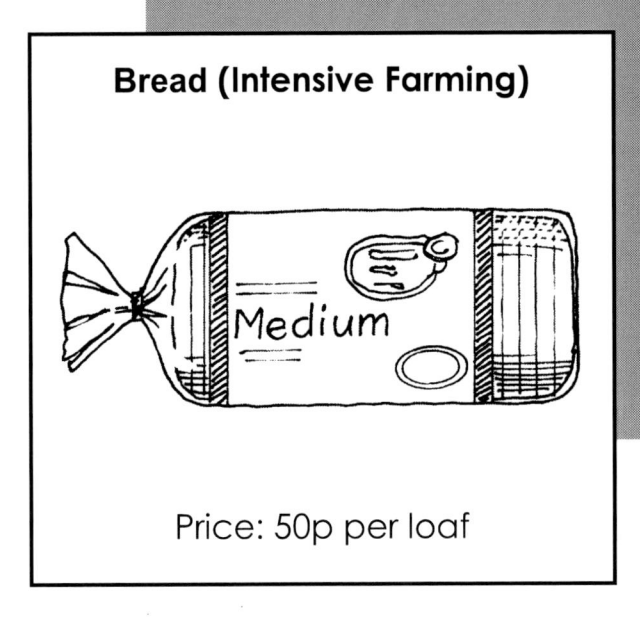

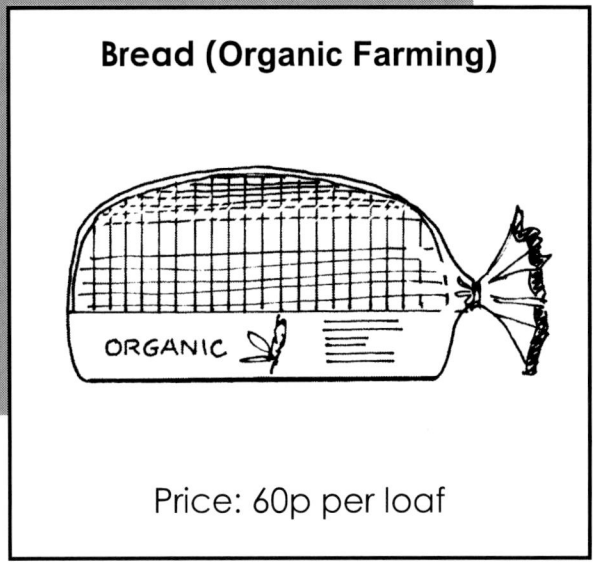

Beef (Intensive Farming)
Price: £14.99 per kg

Beef (Organic Farming)
Price: £25.99 per kg

Bread (Intensive Farming)
Price: 50p per loaf

Bread (Organic Farming)
Price: 60p per loaf

Discuss
A. Why is organic food more expensive than non-organic food?
B. Why is meat so much more expensive than bread?
C. Why do some people prefer to eat organic food?

Be prepared to feed back your thoughts to the rest of the group.

Tip: Write down some bulleted points.

INTERDEPENDENCE & ADAPTATION
WHERE LIMPETS LIVE

4

KS4 NATIONAL CURRICULUM HSW LINK

1. Data, evidence, theories and explanations
   a. how scientific data can be collected and analysed
   b. how interpretation of data, using creative thought, provides evidence to test ideas and develop theories

RESOURCES:
Task Sheet 4, enough for one between two learners.

TIME:
15 minutes for the activity, 10 minutes for class discussion.

NOTES

- This activity is suitable for use as an extended Starter or Plenary, or a Main or Homework activity.
- The task questions get increasingly demanding so, to differentiate, it may be useful to give different groups different questions to discuss and feed back on.

SUGGESTED ANSWERS

A. Identify the independent and the dependent variable.
   Dependent variable: number of limpets per square metre.
   Independent variable: distance from sea.

B. Explain why the boxes with the tide information and rocky shore information are important.
   The tide information tells you when the limpets are covered by the sea and when they are not. The rocky shore information tells you where the rocks are and where a rock pool is.

C. Explain why the number of limpets on the flat rocks before the rock pool is greater than the number of limpets on the flat rocks after the rock pool.
   The limpets before the rock pool are regularly covered by the high tide, so they can feed. The limpets above the rock pool are only covered occasionally by very high tides, so are unable to feed as often.

EXTENSION SUGGESTION

Suggest further research that could be done to study the distribution of limpets.

INDEPENDENCE & ADAPTATION
WHERE LIMPETS LIVE

4

The graph below shows the distribution of limpets on rocks up a rocky shore.

Limpets are small molluscs that attach themselves firmly to rocks when the tide goes out. When the sea covers them, they move over the rocks, eating algae.

The graph below shows the distribution of limpets along a transect up the rocky shore. It also shows the limits of the low tide, high tide and very high tides.

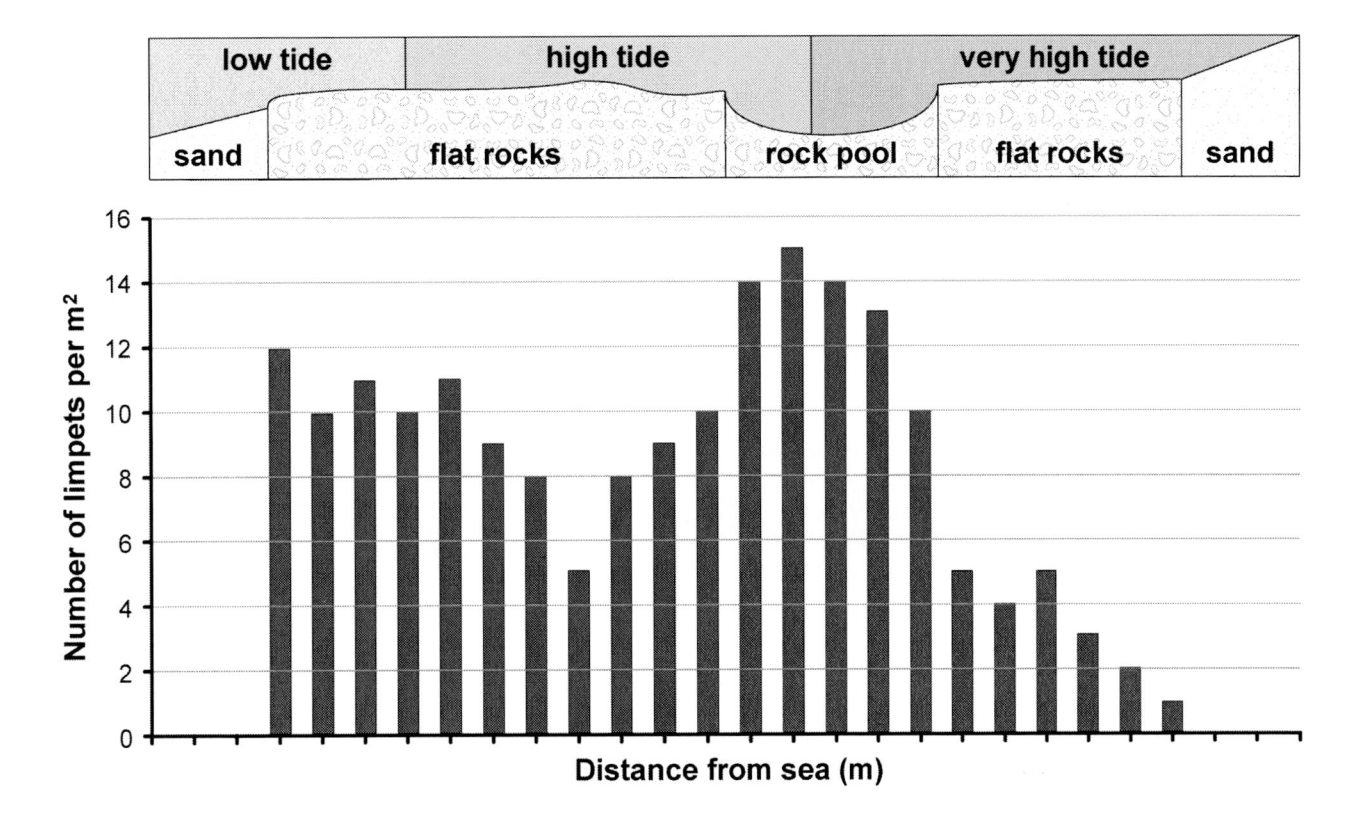

TASK

A. Identify the independent and the dependent variable.
B. Explain why the boxes with the tide information and rock shore information are important.
C. Explain why the number of limpets on the flat rocks before the rock pool is greater than the number of limpets on the flat rocks after the rock pool.

INTERDEPENDENCE & ADAPTATION
FOOD CHAIN CALCULATIONS

5

KS4 NATIONAL CURRICULUM HSW LINK

2. Practical and enquiry skills
   a. plan to test a scientific idea, answer a scientific question or solve a scientific problem
   b. collect data from primary or secondary sources, including using ICT sources and tools
   d. evaluate methods of collection of data and consider their validity and reliability as evidence

RESOURCES:
Task Sheet 5, enough for one per learner or one between two learners.

TIME:
15 minutes for the activity, 10 minutes for class discussion.

NOTES

- This activity is suitable for use as an extended Starter or Plenary, or Main or Homework activity.
- The purpose of the activity is not to actually find out the answer, but to plan to use secondary resources to answer a scientific question.
- Learners must be aware of these key words/concepts before attempting the task: food chains and energy, using secondary resources, hectares.

SUGGESTED ANSWERS

A. Decide on the scientific question you want to answer.
   Based on the instructions on the Task Sheet, for example: Can one hectare of beef cattle feed the same number of people as a hectare of wheat?

B. Make a prediction, using scientific knowledge and understanding.
   A prediction should be made based on knowledge and understanding of energy flow through food chains and pyramids.

C. Write down what you would do to research this question, e.g. books you could use, search engines you could try, the key words you would search for.
   Learners should state how they would search a library catalogue, which search engine they would use and the key words they would try to search.

D. How will you ensure that the information you get is reliable? *Learners should try to find more than one source of information and beware of bias.*

For those who cannot wait to know, my rough calculations were that one hectare can yield about 7,500kg of wheat (which can make about 11,500 loaves) or 110kg of beef (which can make about 440 x 250g beef burgers).

EXTENSION SUGGESTION

Carry out the investigation.

INTERDEPENDENCE & ADAPTATION
FOOD CHAIN CALCULATIONS

5

Consider these two food chains:

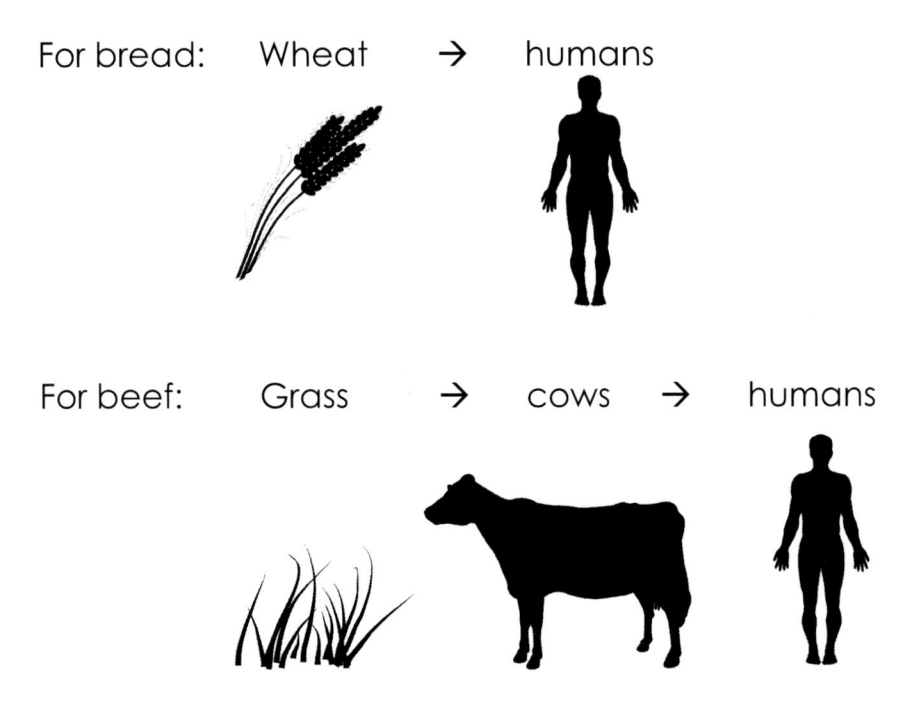

For bread: Wheat   →   humans

For beef: Grass   →   cows   →   humans

This task will help you think about how you can use secondary resources to ask and answer scientific questions.

TASK

Find out whether a hectare of land for beef cattle can feed the same number of people as a hectare of wheat.

Find out how many loaves of bread could be produced from one hectare of wheat.

Find out how many beef burgers could be made from one hectare of beef cattle.

Discuss
A. Decide on the scientific question you want to answer.
B. Make a prediction, using scientific knowledge and understanding.
C. Write down what you would do to research this question, e.g. books you could use, search engines you could try, the key words you would search for.
D. How will you ensure that the information you get is reliable?

From your discussion, write a plan to answer these questions.

INTERDEPENDENCE & ADAPTATION
INTENSIVE FARMING: SOCIAL, ECONOMIC AND ENVIRONMENTAL ISSUES

6

KS4 NATIONAL CURRICULUM HSW LINK

4. Applications and implications of science
   b. to consider how and why decisions about science and technology are made, including those that raise ethical issues, and about the social, economic and environmental effects of such decisions

RESOURCES:
Task Sheet 6 cut into 6 cards, with instructions. Enough for one between two learners.

TIME:
10-15 minutes for the activity, 10 minutes for class discussion.

NOTES

- This activity is suitable for use as a Starter, Main activity or Plenary.
- This can be used as an introduction to the issues of intensive farming or as a consolidation activity.
- Learners must be aware of these key words/concepts before attempting the task: social, economic and environmental.

SUGGESTED ANSWERS

1. Large numbers of animals can be grown in a relatively small space, compared with growing the animals in the field. Farmer   (Ec/Env)
2. The animals are kept in optimum conditions for growth. They are kept warm and dry and given high protein food to help them grow quickly. Farm Hand   (Ec)
3. I am able to ensure that each animal is given antibiotics to prevent disease spreading. This means that very few animals get a disease. Farm Vet   (Ec)
4. The factory farm ensures that there is enough quality meat produced quickly and efficiently to meet the demand of the consumer. Factory Farm Representative   (S/Ec)
5. The animals are kept in cramped conditions, where they cannot move. Some become lame because they grow so fast they cannot support their own bodies. The animals are unable to behave naturally. Animal Welfare Representative   (S)
6. The huge amount of excrement produced by the cows is washed away each day into the ground, contaminating the rivers with antibiotics and growth hormones. Local Resident   (Env/S)

EXTENSION SUGGESTION

Identify the ethical issues associated with intensive farming of livestock.

INTERDEPENDENCE & ADAPTATION
INTENSIVE FARMING: SOCIAL, ECONOMIC AND ENVIRONMENTAL ISSUES

6

TASK

Discuss each statement card and decide whether it is a social, economic or environmental argument about intensive farming.

- **Social:** To do with people's lives and the effect on running a society.
- **Economic:** To do with money, either making money or keeping costs down.
- **Environmental:** To do with keeping our environment unpolluted.

"Large numbers of animals can be grown in a relatively small space, compared with growing the animals in the field." **Farmer**	"The animals are kept in optimum conditions for growth. They are kept warm and dry and given high protein food to help them grow quickly." **Farm Hand**
"I am able to ensure that each animal is given antibiotics to prevent disease spreading. This means that very few animals get a disease." **Farm Vet**	"The factory farm ensures that there is enough quality meat produced quickly and efficiently to meet the demand of the consumer." **Factory Farm Representative**
"The animals are kept in cramped conditions, where they cannot move. Some become lame because they grow so fast they cannot support their own bodies. The animals are unable to behave naturally." **Animal Welfare Representative**	"The huge amount of excrement produced by the cows is washed away each day into the ground, contaminating the rivers with antibiotics and growth hormones." **Local Resident**

INTERDEPENDENCE & ADAPTATION
ORGANIC FARMING: BENEFITS, DRAWBACKS AND RISKS

7

KS4 NATIONAL CURRICULUM HSW LINK

4. Applications and implications of science
   a. about the use of contemporary scientific and technological developments and their benefits, drawbacks and risks

RESOURCES:
Task Sheet 7 cut into 8 cards, with instructions. Enough for one between two or four learners.

TIME:
10 minutes for the activity, 10 minutes for class discussion.

NOTES

- This activity is suitable for use as an extended Starter or Plenary or a Main activity.
- Learners must be aware of these key words/concepts before attempting the task: benefit, drawback and risk.

SUGGESTED ANSWERS

Learners will hopefully recognise that some benefits could also be drawbacks to some people. Below is the intention of each statement: *Benefits: B, Drawbacks: D, Risks: R.*

1. Chemical pesticides are not used, nor are antibiotics or growth hormones. *(B)*
2. Organically grown fruit and vegetables are often odd shapes and have blemishes on them. *(D)*
3. Organically grown food is more expensive than conventionally farmed food. *(D)*
4. Organic crops have a higher chance of failure if a disease or pest infestation occurs. *(R/D)*
5. Organic farms require more space to rear livestock than conventional farms. *(D)*
6. Antibiotics are only used to treat animals when they are sick. *(B)*
7. Some studies have shown organic vegetables to have more vitamins and minerals in them compared to conventionally grown crops. Other studies have shown there to be no difference in nutritional content. *(B)*
8. Since organic crops are fertilised with animal manure, they have to be thoroughly washed before use to avoid gastro-intestinal infections. *(D/R)*

EXTENSION SUGGESTION

List any other benefits, drawbacks or risks you can think of.

INTERDEPENDENCE & ADAPTATION
ORGANIC FARMING: BENEFITS, DRAWBACKS AND RISKS

7

TASK

Discuss each statement card and decide whether it is a benefit, drawback or a risk of organic farming.

- A **benefit** is something that generally has a good effect on people.
- A **drawback** is something that is a hindrance or is the 'downside'.
- A **risk** is a possible danger or source of harm.

Chemical pesticides are not used, nor are antibiotics or growth hormones.	Organically grown fruit and vegetables are often unusual shapes and have blemishes on them.
Organically grown food is more expensive than conventionally farmed food.	Organic crops have a higher chance of failure if a disease or pest infestation occurs.
Organic farms require more space to rear livestock than conventional farms.	Antibiotics are only used to treat animals when they are sick.
Some studies have shown organic vegetables to have more vitamins and minerals in them compared to conventionally grown crops. Other studies have shown there to be no difference in nutritional content.	Since organic crops are fertilised with animal manure, they have to be thoroughly washed before use to avoid gastro-intestinal infections.

VARIATION AND EVOLUTION
HOW WE HAVE GROWN!

8

KS4 NATIONAL CURRICULUM HSW LINK

3. Communication skills
   a. recall, analyse, interpret, apply and question scientific information or ideas
   b. use both qualitative and quantitative approaches
   c. present information, develop an argument and draw a conclusion, using scientific, technical and mathematical language, conventions and symbols, and ICT tools

RESOURCES:
Task Sheet 8, enough for one per learner. Graph paper, pencils and rulers.

TIME:
15 minutes for the activity, 5 minutes for class discussion.

NOTES

- This activity is suitable for use as a Main or Homework activity.
- Learners must be aware of these key words/concepts before attempting the task: presenting information in graphs.

SUGGESTED ANSWERS

A. What is the overall trend of the data?
   Young males are growing taller.
B. What factors may have caused the change in height?
   Improved general health, improved standards of living, improved nutrition in diet.
C. Can you explain why the height of males went down around 1840?
   Economic depression followed by poor diet and disease.

EXTENSION SUGGESTION

Draw a predicted graph for females.

VARIATION AND EVOLUTION
HOW WE HAVE GROWN!

8

The table below shows the average height of 15 year old males in the UK since 1760.

Year	Average height of 15 year old males (cm)
1760	133.4
1780	133.9
1800	131.5
1820	138.2
1840	136.8
1860	139.2
1880	142.8
1900	145.2
1920	152.4
1940	157.2
1960	158.4

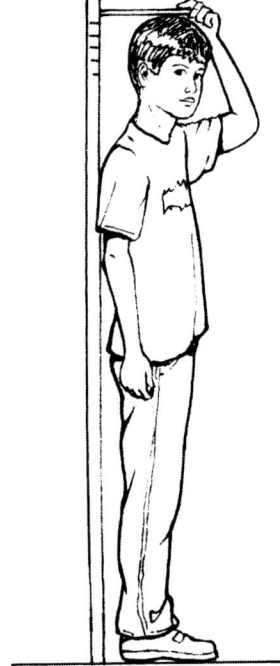

TASK

Plot a graph of this data.

Once you have plotted the graph, answer these questions:

A. What is the overall trend of the data?
B. What factors may have caused the change in height?
C. Can you explain why the height of males went down around 1840?

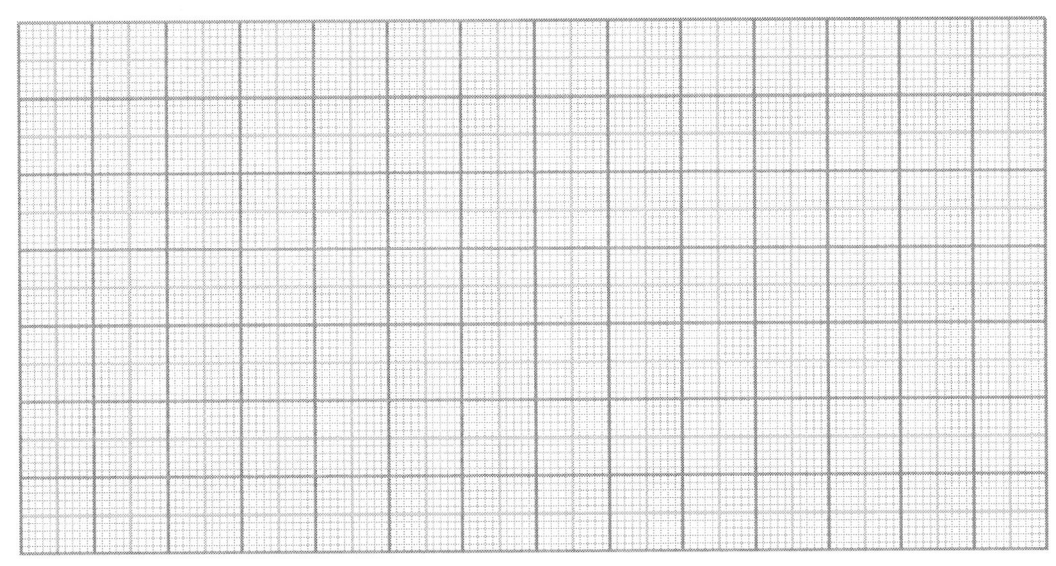

VARIATION AND EVOLUTION
IVY LEAVES

9

KS4 NATIONAL CURRICULUM HSW LINK

1. Data, evidence, theories and explanations
   b. how interpretation of data, using creative thought, provides evidence to test ideas and develop theories
   c. how explanations of many phenomena can be developed using scientific theories, models and ideas

RESOURCES:
Task Sheet 9, enough for one per learner.

TIME:
10 minutes for the activity, 10 minutes for class discussion.

NOTES

- This activity is suitable for use as a Starter, Main activity or Plenary.
- Learners must be aware of these key words/concepts before attempting the task: competition in plants for light, genetic and environmental variation, photosynthesis.

SUGGESTED ANSWERS

Ivy is a common plant that climbs walls and trees. If you measure the leaves at the bottom of the tree, they have a larger surface area than the leaves towards the top of the tree. Why?

1. *The leaves at the bottom are older and therefore bigger than the younger leaves at the top.*
2. *There is less light at the base of the tree, so the leaves at the bottom of the tree grow bigger to absorb more light.*

EXTENSION SUGGESTION

Plan how you would test your prediction.

VARIATION AND EVOLUTION
IVY LEAVES

9

Ivy is a common plant that climbs walls and trees. If you measure the leaves at the bottom of the tree, they have a larger surface area than the leaves towards the top of the tree.

TASK

Discuss why you think this is.

Write some predictions that could be tested scientifically.

VARIATION AND EVOLUTION
DINOSAUR EXTINCTION

10

KS4 NATIONAL CURRICULUM HSW LINK

1. Data, evidence, theories and explanations
   a. how scientific data can be collected and analysed
   b. how interpretation of data, using creative thought, provides evidence to test ideas and develop theories
   c. how explanations of many phenomena can be developed using scientific theories, models and ideas
   d. that there are some questions that science cannot currently answer, and some that science cannot address

RESOURCES:
Task Sheet 10, enough for one per learner.

TIME:
15 minutes for the activity, 10 minutes for class discussion.

NOTES

- This activity is suitable for use as an extended Starter or Plenary, or Main or Homework activity.
- Learners with reading difficulties will need support either by a peer or by the teacher reading through the text.
- Learners must be aware of these key words/concepts before attempting the task: extinction, theories.

SUGGESTED ANSWERS

A. What the two scientists agree on. *The dinosaurs became extinct. The extinction occurred at the Cretaceous-Triassic time (65 million years ago). A large number of species became extinct at that time. There is an unusual layer of clay found in rocks, with high amounts of iridium. Change of climate led to extinction.*

B. What the two scientists disagree about. *What caused the extinction: either asteroid impact or increased volcanic activity. From where the iridium originated (either asteroid or Earth's core). Exact percentage of species that became extinct.*

C. The evidence they use in their arguments. *Geological information. Chemical analysis of clay layer in rock. Gas analysis of ice cores, fossil record.*

D. Why it is difficult to be sure exactly why the dinosaurs became extinct. *Incomplete fossil record, no one was there to witness it.*

EXTENSION SUGGESTION

What theories do you have to explain why the dinosaurs became extinct?

VARIATION AND EVOLUTION
DINOSAUR EXTINCTION

10

TASK

Everybody knows that dinosaurs became extinct but, because they died out millions of years ago, it is difficult to know exactly why they died out.

Read the views of these two scientists. Then complete the task below:

Prof. May B. Wright

Despite there being many theories of the cause of dinosaur extinction, the one that there is most evidence for is the asteroid theory. Many scientists around the world have found a layer of clay in rocks that indicated that there was a huge impact on Earth at the time between the Cretaceous and Triassic periods. The clay contains unusually high amounts of the element iridium. Asteroids contain much higher levels of iridium than Earth rock. The asteroid impact would have covered a great proportion of the Earth with a large layer of dust. The debris from the collision in the atmosphere would have shut out sunlight, causing plants to die and so the dinosaurs that depended on the plants to starve. The temperature would have decreased and the dinosaurs would have died of the cold as well as starvation. Not only the dinosaurs died out but, also, the fossil record shows that 70% of all species became extinct at this time.

Dr. Ivor Goodguess

The dinosaur extinction was most likely caused by a huge rise in volcanic activity followed by climate change. This change did not just kill off the dinosaurs, but many other species. Estimates suggest that between 60 and 80 percent of Earth's species died at this time in a mass extinction. There is an unusual layer of clay found in rocks, which formed about 65 million years ago. This coincides with the mass extinction in the fossil record. This layer of clay contains iridium, an element that is found in the Earth's core. Many years before the extinction, there is evidence of an increased amount of volcanic activity. The dust and gases that were produced contained gases from the Earth's core. The volcanic activity caused the atmosphere to change. First, the concentration of carbon dioxide was twelve times higher than it is today, which we have found from analysing gas trapped in ice cores. Also, the average temperature of the Earth increased by about 10°C. These rapid changes in climate and atmosphere caused many land living species to become extinct.

Discuss and decide on:
A. What the two scientists agree on.
B. What the two scientists disagree about.
C. The evidence they use in their arguments.
D. Why it is difficult to be sure exactly why the dinosaurs became extinct.

KS4 BIOLOGY HOW SCIENCE WORKS: COPYMASTER
Biology HSW © Badger Publishing Ltd

VARIATION AND EVOLUTION
FOSSIL EVIDENCE

11

KS4 NATIONAL CURRICULUM HSW LINK

1. Data, evidence, theories and explanations
   a. how scientific data can be collected and analysed
   b. how interpretation of data, using creative thought, provides evidence to test ideas and develop theories
   c. how explanations of many phenomena can be developed using scientific theories, models and ideas
   d. that there are some questions that science cannot currently answer, and some that science cannot address

RESOURCES:
Task Sheet 11 cut into 9 cards, with instructions. Enough for one between two or four learners.

TIME:
5-10 minutes for the activity, 10 minutes for class discussion.

NOTES

- This activity is suitable for use as a Starter, Main activity or Plenary.
- Learners must be aware of these key words/concepts before attempting the task: types of evidence, fossilisation.

SUGGESTED ANSWERS

1. The dinosaur was 1.2 metres tall. — (Direct)
2. The scales of the dinosaur were a grey-green colour. — (No evidence)
3. The dinosaur hunted in packs. — (No evidence)
4. The dinosaur was a carnivore. — (Direct)
5. The dinosaur could make loud chirping sounds. — (No evidence)
6. The dinosaur used its tail for balance. — (Indirect)
7. The dinosaur is of a new species called Theropods. — (Indirect)
8. The dinosaur lived over 150 million years ago. — (Indirect)
9. The dinosaur walked on two legs. — (Direct)

EXTENSION SUGGESTION

For the 'no evidence' cards, what evidence would scientists need to make the claim?

KS4 Biology How Science Works: Teacher notes

Variation and Evolution
Fossil Evidence

11

Fossil hunters found a complete fossil skeleton of a new type of dinosaur in the Isle of Wight, UK. The palaeontologists were able to draw a number of conclusions from the find. Palaeontologists draw conclusions from a variety of sources of evidence, not just the bones.

Task

Sort these cards into three groups.
- **Direct evidence** from the fossilised bones.
- **Indirect evidence** from the fossilised bones.
- **No evidence** from the fossilised bones.

How do you think palaeontologists draw conclusions that are not directly based on the fossilised bones?

The dinosaur was 1.2 metres tall.	The scales of the dinosaur were a grey-green colour.	The dinosaur hunted in packs.
The dinosaur was a carnivore.	The dinosaur could make loud chirping sounds.	The dinosaur used its tail for balance.
The dinosaur is of a new species called Theropods.	The dinosaur lived over 150 million years ago.	The dinosaur walked on two legs.

KS4 Biology How Science Works: Copymaster
Biology HSW © Badger Publishing Ltd

VARIATION AND EVOLUTION
EVOLUTION OF THE HORSE

12

KS4 NATIONAL CURRICULUM HSW LINK

3. Communication skills
   a. recall, analyse, interpret, apply and question scientific information or ideas
   b. use both qualitative and quantitative approaches

RESOURCES:
Task Sheet 12 cut up into 15 cards, with instructions. Enough for one set between 2-4 learners.

TIME:
10-15 minutes, 10 minutes for class discussion.

NOTES

- This activity is most suitable as a Main or Homework activity.
- Some learners may need support with the more unusual words in this activity.
- Learners must be aware of these key words/concepts before attempting the task: evolution, omnivore.

SUGGESTED ANSWERS

52-45 million years ago (Eocene)	*Hyracotherium* (*Eohippus*) Height 0.4m	Habitat is covered in small trees and thickets, providing lots of cover. The ground is marshy. This horse has four toes on the front feet, three toes on the rear feet. Teeth are typical of an omnivore.
37-32 million years ago (Oligocene)	*Mesohippus* (Middle Horse) Height 0.6m	Large grasslands are developing, offering few places to hide. The climate is becoming drier. Front foot now has three toes. Molars develop into battery of teeth for grinding.
17-11 million years ago (Miocene)	*Merychippus* Height 1.0m	The middle (third) toe has developed a large convex 'hoof' and the lower leg bones have fused, enabling the horse to run very fast over hard ground. Begins to eat grass predominantly.
12-6 million years ago (Pliocene)	*Pliohippus* Height 1.2m	The side toes are lost and horse has become 'one toed'. Side ligaments develop around the toe to stabilise the central toe when running.
5 million years ago to present (Pleistocene)	*Equus* (Modern horse) Height 1.6m	One toed with ligaments that prevent the hoof from twisting. Long neck and long legs with fused leg bones. Has high crowned, grazing teeth. Body is stocky with stiff, upright mane.

EXTENSION SUGGESTION

How did the horse's evolution respond to its changing environment?

Variation and Evolution
Evolution of the Horse

12

Task

Arrange the timeline in chronological order. Match the horse with when it evolved and how its leg bones and teeth developed. Discuss the ways in which horses, and in particular the leg bones and teeth, have adapted to the changing environment.

12-6 million years ago (Pliocene)	*Equus* (Modern horse) Height 1.6m	Large grasslands are developing, offering few places to hide. The climate is becoming drier. Front foot now has three toes. Molars develop into battery of teeth for grinding.
17-11 million years ago (Miocene)	*Hyracotherium* (*Eohippus*) Height 0.4m	The side toes are lost and horse has become 'one toed'. Side ligaments develop around the toe to stabilise the central toe when running.
52-45 million years ago (Eocene)	*Merychippus* Height 1.0m	One toed with ligaments that prevent the hoof from twisting. Long neck and long legs with fused leg bones. Has high crowned, grazing teeth. Body is stocky with stiff, upright mane.
37-32 million years ago (Oligocene)	*Mesohippus* (Middle Horse) Height 0.6m	Habitat is covered in small trees and thickets, providing lots of cover. The ground is marshy. This horse has four toes on the front feet, three toes on the rear feet. Teeth are typical of an omnivore.
5 million years ago to present (Pleistocene)	*Pliohippus* Height 1.2m	The middle (third) toe has developed a large convex 'hoof' and the lower leg bones have fused, enabling the horse to run very fast over hard ground. Begins to eat grass predominantly.

KS4 Biology How Science Works: Copymaster
Biology HSW © Badger Publishing Ltd

VARIATION AND EVOLUTION
How to Clone

13

KS4 NATIONAL CURRICULUM HSW LINK

3. Communication skills
   a. recall, analyse, interpret, apply and question scientific information or ideas
   c. present information, develop an argument and draw a conclusion, using scientific, technical and mathematical language, conventions and symbols, and ICT tools

RESOURCES:
Task Sheet 13 cut into 12 cards, with instructions. Enough for one set between two learners.

TIME:
10 minutes for the activity, 10 minutes for class discussion.

NOTES

- This activity could be used as a Main activity, Plenary or a Homework activity.
- This can be used as a stimulus activity when introducing learners to the topic of cloning.
- Learners must be aware of these key words before attempting the task: clone, de-nucleated, donor, DNA, surrogate.
- For more able learners, the pictures could be removed from the statements and ordered separately.

SUGGESTED ANSWERS

Egg is removed from egg donor sheep. The nucleus is removed to form a 'de-nucleated' egg.	
A cell is removed from the mammary gland of the DNA donor sheep. This cell will donate the DNA to the cloned sheep.	
The 'de-nucleated' egg and the donated DNA are fused together using an electric shock.	
The fused cell is given another electric shock and begins to divide to form an embryo.	
The embryo is implanted into the uterus of a surrogate sheep.	
The surrogate sheep gives birth to a lamb. The lamb is a clone of the DNA donor sheep.	

EXTENSION SUGGESTION

Suggest any benefits or drawbacks of cloning.

KS4 BIOLOGY HOW SCIENCE WORKS: TEACHER NOTES

Variation and Evolution
How to Clone

13

TASK

Read through the statements on the cards. Place the cards in the correct order to show how a sheep can be cloned.

Egg is removed from egg donor sheep. The nucleus is removed to form a 'de-nucleated' egg.	
The 'de-nucleated' egg and the donated DNA are fused together using an electric shock.	A — Body cell taken from Sheep A — DNA extracted
The surrogate sheep gives birth to a lamb. The lamb is a clone of the DNA donor sheep.	
The fused cell is given another electric shock and begins to divide to form an embryo.	B — Egg cell taken from Sheep B — Nucleus removed
A cell is removed from the mammary gland of the DNA donor sheep. This cell will donate the DNA to the cloned sheep.	Lamb is clone of Sheep A
The embryo is implanted into the uterus of a surrogate sheep.	DNA extracted — Nucleus removed — DNA from Sheep A fused with egg cell from Sheep B

KS4 Biology How Science Works: Copymaster
Biology HSW © Badger Publishing Ltd

VARIATION AND EVOLUTION
DESIGNER BABIES

14

KS4 NATIONAL CURRICULUM HSW LINK

4. Applications and implications of science
   a. about the use of contemporary scientific and technological developments and their benefits, drawbacks and risks
   b. to consider how and why decisions about science and technology are made, including those that raise ethical issues, and about the social, economic and environmental effects of such decisions
   c. how uncertainties in scientific knowledge and scientific ideas change over time and about the role of the scientific community in validating these changes

RESOURCES:
Task Sheet 14, enough for one between two learners.

TIME:
5-10 minutes for the activity, 5-10 minutes for class discussion.

NOTES

- This activity is suitable for use as a Starter or Plenary, or Main or Homework activity.
- Learners must be aware of these key words/concepts before attempting the task: human genome, genes, basic inheritance, process of embryo screening.

SUGGESTED ANSWERS

Learners will come up with their own arguments but it should be noted that, in the UK, it is illegal to 'design babies' unless there is a strong medical argument for it, such as 'saving a sibling'.

EXTENSION SUGGESTION

If it was up to you, which characteristics would you let parents select and why?

VARIATION AND EVOLUTION
DESIGNER BABIES

14

TASK

As scientists understand the human genome, more genes are isolated for more characteristics.

Discuss each of these situations and whether you think parents should be able to select these characteristics for their babies.

Should we be able to:

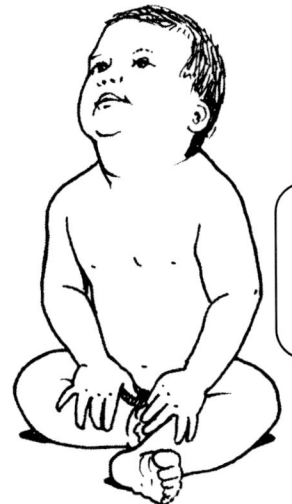

- select whether to have a boy or girl?

- choose to have a girl because a boy is likely to have haemophilia?

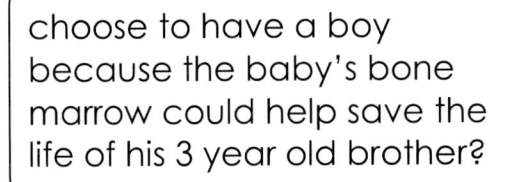

- choose to have a boy because the baby's bone marrow could help save the life of his 3 year old brother?

- select the eye colour or hair colour of the baby?

- choose to have a gene added that would improve the baby's intelligence?

KS4 Biology How Science Works: Copymaster
Biology HSW © Badger Publishing Ltd

Variation and Evolution
Evolution Timeline

15

KS4 National Curriculum HSW link

4. Applications and implications of science
   c. how uncertainties in scientific knowledge and scientific ideas change over time and about the role of the scientific community in validating these changes

Resources:
Task Sheet 15, cut into 15 cards, with instructions. Enough sets for one per learner.

Time:
20-30 minutes for the activity, 10 minutes for class discussion.

Notes

- This activity is suitable for use as a Main or Homework activity.
- Learners must be aware of these key words/concepts before attempting the task: evolution and natural selection, how scientific ideas change.

Suggested answers

Most people believed in creationism, that a god made all the species of plants and animals, and that these species do not change over time.
Charles Cuvier, in 1796, linked fossils of mammoths and mastodons to living elephants, providing evidence that some species became extinct.
Jean-Baptiste Lamarck, in 1809, noted that animals were adapted to their environment and suggested that species may change over time.
1830s. Charles Lyell wrote a book explaining that the geological changes took place over millions of years.
Charles Darwin took a voyage on HMS Beagle in the 1830s. While on the Galapagos Islands, Darwin came up with his idea of natural selection when studying the species of finch there.
In 1859, Charles Darwin published *The Origin of Species*. He explained that all organisms evolved from a common ancestor through a process called natural selection.
Gregor Mendel's work on the laws of inheritance in peas was found in 1900 and provided evidence for Darwin's theory of evolution.
In 1953, Watson and Crick, based on Rosalind Franklin's work, discovered the structure of DNA. From this, the concept of the 'gene' was developed.
Evolution through natural selection could be explained through the gene. In the 1960s, scientists were discussing if evolution happened very gradually or quickly followed by long periods of stable species.
Scientists by-pass natural selection, by isolating genes using biotechnology. These genes can be used to genetically modify organisms.

Extension suggestion

Predict what we might know and be able to do in the future.

VARIATION AND EVOLUTION
EVOLUTION TIMELINE

15

TASK

Cut out the timeline, the statements and the images. Read each statement carefully and place it on the timeline. Match the images to the timeline to illustrate it.

Gregor Mendel's work on the laws of inheritance in peas was found in 1900 and provided evidence for Darwin's theory of evolution.	
In 1859, Charles Darwin published *The Origin of Species*. He explained that all organisms evolved from a common ancestor through a process called natural selection.	
1830s. Charles Lyell wrote a book explaining that the geological changes took place over millions of years.	
Scientists by-pass natural selection, by isolating genes using biotechnology. These genes can be used to genetically modify organisms.	
Charles Cuvier, in 1796, linked fossils of mammoths and mastodons to living elephants, providing evidence that some species became extinct.	
Charles Darwin took a voyage on HMS Beagle in the 1830s. While on the Galapagos Islands, Darwin came up with his idea of natural selection when studying the species of finch there.	
Most people believed in creationism, that a god made all the species of plants and animals, and that these species do not change over time.	
In 1953, Watson and Crick, based on Rosalind Franklin's work, discovered the structure of DNA. From this, the concept of the 'gene' was developed.	
Evolution through natural selection could be explained through the gene. In the 1960s, scientists were discussing if evolution happened very gradually or quickly followed by long periods of stable species.	
Jean-Baptiste Lamarck, in 1809, noted that animals were adapted to their environment and suggested that species may change over time.	

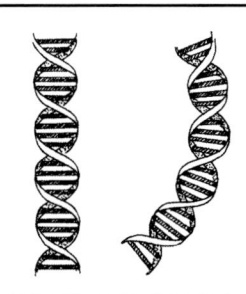

KS4 BIOLOGY HOW SCIENCE WORKS: COPYMASTER

Biology HSW © Badger Publishing Ltd

GENES AND GENETICS
TESTING ANTIBIOTICS

16

KS4 NATIONAL CURRICULUM HSW LINK

3. Communication skills
   a. recall, analyse, interpret, apply and question scientific information or ideas
   b. use both qualitative and quantitative approaches
   c. present information, develop an argument and draw a conclusion, using scientific, technical and mathematical language, conventions and symbols, and ICT tools

RESOURCES:
Task Sheet 16, enough for one per learner or one between two learners.

TIME:
10 minutes for the activity, 5 minutes for class discussion.

NOTES

- This activity is suitable for use as a Starter or Plenary, or Main or Homework activity.
- Learners must be aware of these key words/concepts before attempting the task: Petri dish, aseptic, antibiotics, *E. coli*.

SUGGESTED ANSWERS

A. What is the purpose of the control disc 'A'?
   To see the effect of no antibiotic.
B. Put the antibiotics in order of effectiveness against this strain of the bacteria *E.coli*. (Most effective first). *Penicillin (B), Ampicillin (C) then Streptomycin (D).*
C. To which antibiotic is this strain of *E.coli* resistant?
   Streptomycin (same as the control A).
D. How could bacteriologists take quantitative measurements?
   Measure the diameter of the inhibited zone in mm.
E. How could bacteriologists make sure that these results were reliable?
   Repeat in several Petri dishes and take the average inhibition diameter.

EXTENSION SUGGESTION

List some situations when doctors would use this technique.

KS4 BIOLOGY HOW SCIENCE WORKS: Teacher notes

GENES AND GENETICS
TESTING ANTIBIOTICS

16

On agar jelly in a Petri dish, a bacteria is grown on the surface.

Small disks of filter paper that have been soaked in antibiotic solution can be placed on the surface of the bacteria.

After 24 hours, scientists can see if the antibiotic is effective at killing the bacteria by looking at the bacteria around the antibiotic-soaked disks. If the antibiotic kills the bacteria, it can be seen because around the disc will be a clear space where the bacteria have died.

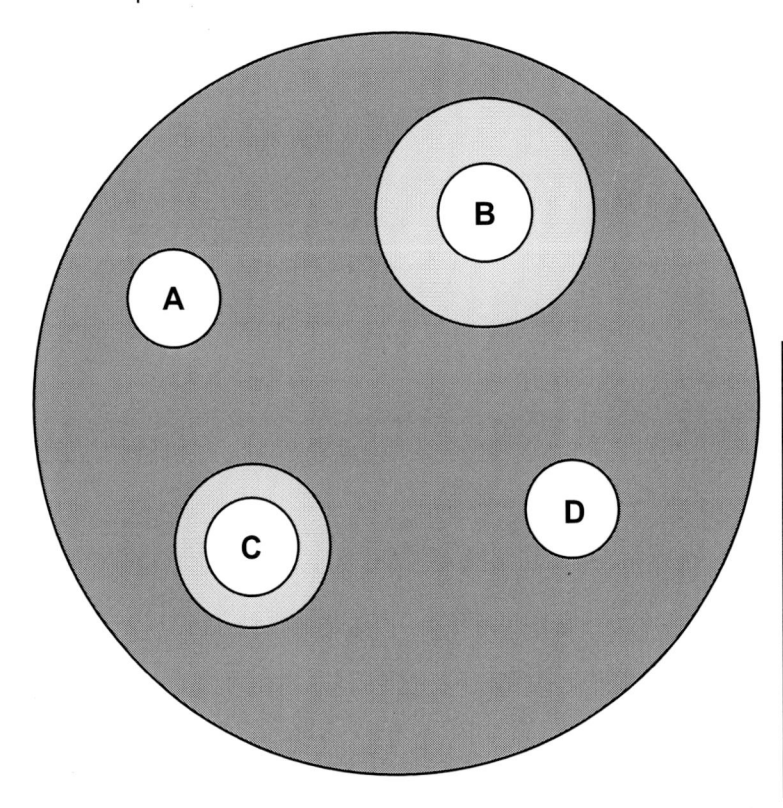

Bacteria *E.coli* ▮

No *E.coli.* ▯

A. Control (distilled, aseptic water)
B. Penicillin
C. Ampicillin
D. Streptomycin

TASK

Look at the Petri dish above.

Using the information above and the key, discuss the following questions:

A. What is the purpose of the control disc 'A'?
B. Put the antibiotics in order of effectiveness against this strain of the bacteria *E.coli*. (Most effective first.)
C. To which antibiotic is this strain of *E.coli* resistant?
D. How could bacteriologists take quantitative measurements?
E. How could bacteriologists make sure that these results were reliable?

GENES AND GENETICS
GENETIC MODIFICATION OF CROPS: FACT OR OPINION?

17

KS4 NATIONAL CURRICULUM HSW LINK

4. *Applications and implications of science*
   c. *how uncertainties in scientific knowledge and scientific ideas change over time and about the role of the scientific community in validating these changes*

RESOURCES:
Task Sheet 17 cut into 9 cards, with instructions. Enough for one set between 2-4 learners.

TIME:
10 minutes plus discussion time.

NOTES

- This activity is suitable as a Starter, Main activity or Plenary.
- You may need to define the terms: genetic modification, yield, contaminate – if learners have not encountered them.

SUGGESTED ANSWERS

1. Genetic modification will reduce the amount of insecticides used on crops. *(Fact/For)*
2. Traditional breeding methods are safer than genetic modification. *(Against)*
3. Genetic modification is the changing of an organism's DNA. *(Fact)*
4. The potential benefits of genetic modification of crops are enormous. *(For)*
5. Genes may escape from a GM crop and contaminate wild species. *(Fact)*
6. The long term effects of GM crops on the environment and health are unknown at the moment. *(Fact/Against)*
7. Genetically modified crops and foods are harmful to our health. *(Against)*
8. Genetic modification of crops will result in increased yields and increased profits for farmers. *(Fact/For)*
9. Genetic modification of crops is unnatural and tinkering with nature. *(Against)*

EXTENSION SUGGESTION

List any questions you have about the statements.

KS4 BIOLOGY HOW SCIENCE WORKS: TEACHER NOTES

Genes and Genetics
Genetic Modification of Crops: Fact or Opinion?

17

Task

Discuss each statement card and decide whether it is a fact, an opinion for or an opinion against genetic modification of crops. Place the cards into three piles:

- **Facts** about genetic modification.
- **Opinions for** genetic modification.
- **Opinions against** genetic modification.

A fact is something that has been proved true. An opinion is something that someone feels or believes, whether or not it is true.

Genetic modification will reduce the amount of insecticides used on crops.	Traditional breeding methods are safer than genetic modification.	Genetic modification is the changing of an organism's DNA.
The potential benefits of genetic modification of crops are enormous.	Genes may escape from a GM crop and contaminate wild species.	The long term effects of GM crops on the environment and health are unknown at the moment.
Genetically modified crops and foods are harmful to our health.	Genetic modification of crops will result in increased yields and increased profits for farmers.	Genetic modification of crops is unnatural and tinkering with nature.

GENES AND GENETICS
INDUSTRIAL FERMENTERS

18

KS4 NATIONAL CURRICULUM HSW LINK

2. Practical and enquiry skills
   a. plan to test a scientific idea, answer a scientific question or solve a scientific problem
   b. collect data from primary or secondary sources, including using ICT sources and tools

RESOURCES:
Task Sheet 18, enough for one between two learners, OHT of Task Sheet with fermenter information.

TIME:
10 minutes, 5 minutes for class discussion.

NOTES

- This activity is suitable as a Starter, Main activity or Plenary. It can be used as a stimulus activity when introducing how bacteria are grown in culture to produce microbial products, e.g. insulin, human growth hormone.
- You may need to define the terms: fermentation, microbes, agitation, contaminants, sterilise, bacterial cultures.

SUGGESTED ANSWERS

Design feature	Reason for design feature
Cooling jacket	*To maintain a constant temperature; heat energy is generated through respiration (and through movement of paddles); optimum temperature needs to be maintained to ensure optimum growth of bacteria.*
Air filter	*To sterilise incoming air; to remove contaminants.*
Air pump	*To move the air through the broth so oxygen is evenly distributed; also provides further agitation.*
Air outlet	*To remove waste gases; prevent build up of pressure.*
Temperature recorder	*To record temperature (linked to cooling jacket).*
Harvest line	*To allow bacterial culture/product to be removed.*
Tap	*To regulate removal of bacterial product/culture.*
Paddle wheel	*To provide agitation to evenly distribute gases and nutrients; to prevent clumping of bacteria.*

EXTENSION SUGGESTION

What other conditions may need to be monitored and controlled in the fermenter?

GENES AND GENETICS
INDUSTRIAL FERMENTERS

18

Industrial fermenters, like of the one shown here, are used to grow microbes that make useful products, e.g. beer or insulin.

Conditions inside the fermenter have to be carefully monitored and controlled so that the microbes have the best environment for optimum (maximum) growth and production of the useful substance. This diagram shows the inside of a fermenter, with some of the design features that allow the conditions inside the fermenter to be kept constant and at the correct levels.

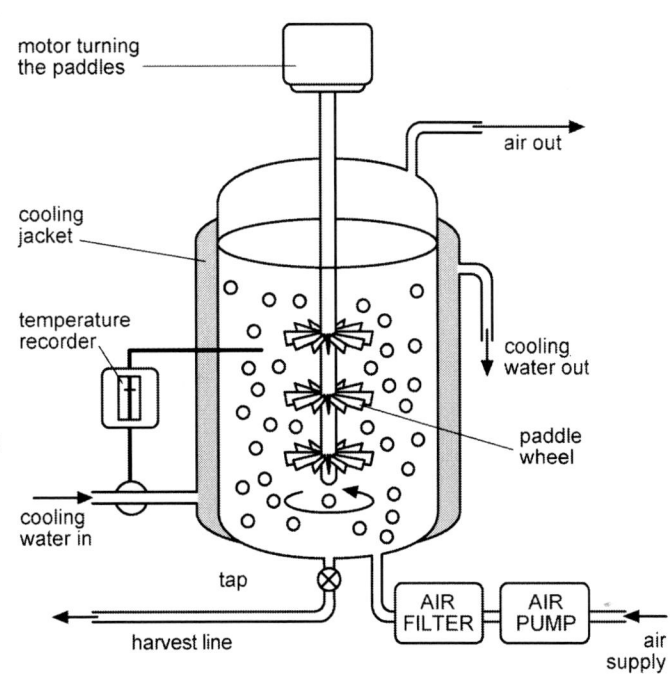

TASK

Match the card for 'reason for design feature' to the correct 'design feature':

Design feature	Reason for design feature
Cooling jacket	To remove waste gases; prevent build up of pressure.
Air filter	To provide agitation to evenly distribute gases and nutrients; to prevent clumping of bacteria.
Air pump	To record temperature (linked to cooling jacket).
Air outlet	To maintain a constant temperature; heat energy is generated through respiration; optimum temperature needs to be maintained to ensure optimum growth of bacteria.
Temperature recorder	To move the air through the broth so oxygen is evenly distributed; also provides further agitation.
Harvest line	To allow bacterial culture/product to be removed.
Tap	To sterilise incoming air; to remove contaminants.
Paddle wheel	To regulate removal of bacterial product/culture.

KS4 Biology How Science Works: Copymaster
Biology HSW © Badger Publishing Ltd

GENES AND GENETICS
HOW GENE SPLICING WORKS

19

KS4 NATIONAL CURRICULUM HSW LINK

2. Practical and enquiry skills
   a. plan to test a scientific idea, answer a scientific question or solve a scientific problem
   b. collect data from primary or secondary sources, including using ICT sources and tools

RESOURCES:
Task Sheet 19 cut up into 6 cards (or 12 for more able learners), with instructions. Enough for one set between 2-4 learners.

TIME:
10 minutes.

NOTES

- This activity is suitable as a Starter, Main activity or Plenary.
- You may need to define the terms: splice, ligase, sticky ends, restriction enzyme – if learners have not encountered them.
- For more able learners, the pictures could be cut off the statements and these also arranged in the correct order.

SUGGESTED ANSWERS

DNA removed from human cell. Human insulin gene 'cut' out of DNA using restriction enzyme.	human insulin gene / section of human DNA
An enzyme cuts open bacterial plasmid.	bacterial plasmid / plasmid cut open with restriction enzyme
Ligase enzyme joins gene to cut ends of plasmid.	ligase enzyme sticks insulin gene into plasmid
Plasmid with human gene is reinserted into bacterium.	plasmid put into bacterium
Bacterium grows and reproduces quickly; each bacterial cell produces insulin using the human insulin gene.	bacteria with insulin gene grown in fermenter
In the correct conditions, the bacteria produce large quantities of human insulin.	insulin is separated off and purified

EXTENSION SUGGESTION

List any new key words you have come across during this activity and find out what they mean.

GENES AND GENETICS
HOW GENE SPLICING WORKS

19

TASK

Sort the cards (and pictures) into the correct order to show how gene splicing works.

An enzyme cuts open bacterial plasmid.	bacterial plasmid — plasmid cut open with restriction enzyme
Bacterium grows and reproduces quickly; each bacterial cell produces insulin using the human insulin gene.	bacteria with insulin gene grown in fermenter
Ligase enzyme joins gene to cut ends of plasmid.	ligase enzyme sticks insulin gene into plasmid
In the correct conditions, the bacteria produce large quantities of human insulin.	insulin is separated off and purified
DNA removed from human cell. Human insulin gene 'cut' out of DNA using restriction enzyme.	human insulin gene — section of human DNA
Plasmid with human gene is reinserted into bacterium.	plasmid put into bacterium

KS4 BIOLOGY HOW SCIENCE WORKS: COPYMASTER
Biology HSW © Badger Publishing Ltd

GENES AND GENETICS
DNA FINGERPRINT ANALYSIS

20

KS4 NATIONAL CURRICULUM HSW LINK

3. Communication skills
   a. recall, analyse, interpret, apply and question scientific information or ideas
   b. use both qualitative and quantitative approaches

RESOURCES:
Task Sheet 20, enough for one between two learners.

TIME:
10 minutes for the activity, 5 minutes for class discussion.

NOTES

- This activity is suitable as a Starter, Main activity or Plenary.
- You may need to define the terms: genetic engineering, DNA, DNA fingerprint, replicate.

SUGGESTED ANSWERS

Person C. This person's DNA fingerprint matches sample X in band widths and spacing.

EXTENSION SUGGESTION

List any new key words you have come across during this activity and find out what they mean.

KS4 BIOLOGY HOW SCIENCE WORKS: TEACHER NOTES

GENES AND GENETICS
DNA FINGERPRINT ANALYSIS

20

DNA fingerprinting started about 20 years ago. The DNA can come from a little bit of saliva, blood, skin, etc. The amount of DNA needed to make a fingerprint is minute – a single human cell will provide enough DNA to make a fingerprint.

The chance of someone having the same DNA as you is minuscule (unless you have an identical twin). DNA fingerprinting can be used to solve crimes and settle paternal suites (finding out who is someone's actual biological father). DNA fingerprinting can also be used to identify genes that cause inherited conditions, e.g. cystic fibrosis, and genes that may increase the chances of developing cancer.

TASK

Look at the DNA fingerprints below. Sample X was found at a crime scene. Suspects A, B, C and D gave samples of their saliva to check if their DNA matched that at the crime scene. Discuss the DNA fingerprints with your partner. Decide which person left the sample at the crime scene. How do you explain your choice?

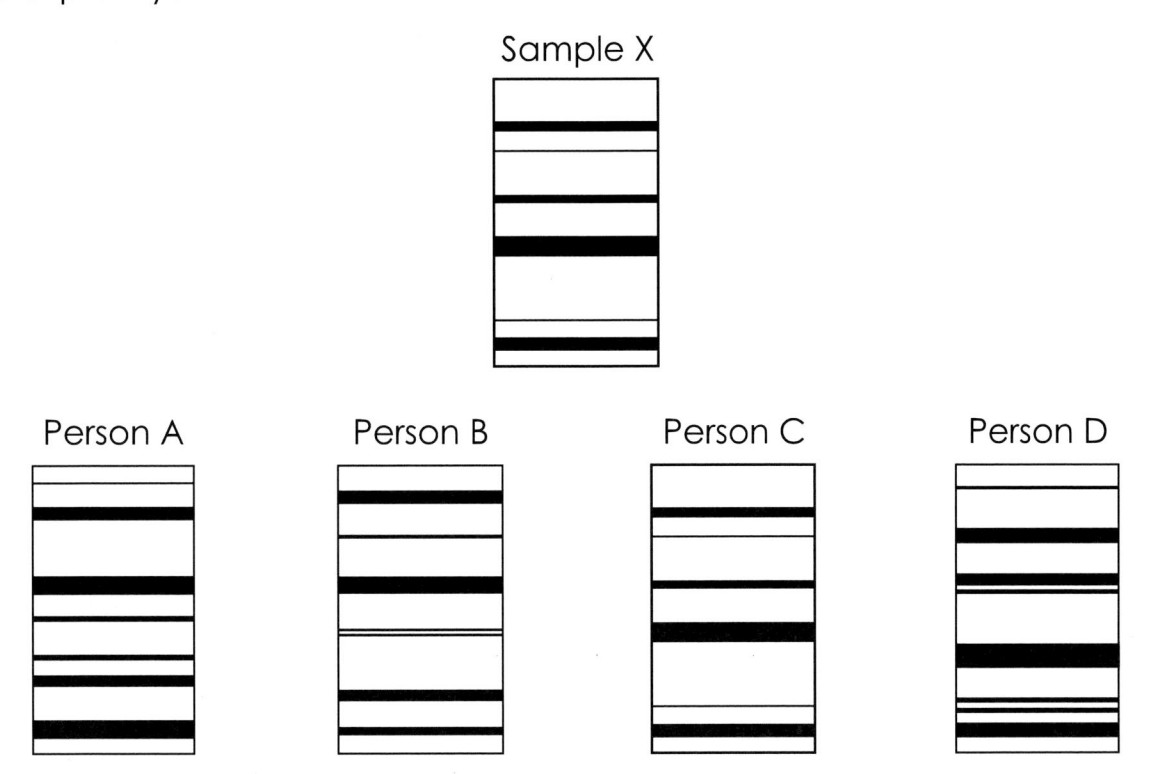

Sample X

Person A   Person B   Person C   Person D

Discuss
Identify the problems with trying to match genetic fingerprints.
How sure would you be that you had got the right person?

KS4 BIOLOGY HOW SCIENCE WORKS: COPYMASTER
Biology HSW © Badger Publishing Ltd

GENES AND GENETICS
GENETICS TIMELINE

21

KS4 NATIONAL CURRICULUM HSW LINK

4. Applications and implications of science
   c. how uncertainties in scientific knowledge and scientific ideas change over time and about the role of the scientific community in validating these changes

RESOURCES:
Task Sheet 21, cut into 16 cards, with instructions. Enough for one per learner.

TIME:
20-30 minutes for the activity, 10 minutes for class discussion.

NOTES

- This activity is suitable for use as a Main or Homework activity.
- Learners must be aware of these key words/concepts before attempting the task: DNA and genetics, how scientific ideas change.

SUGGESTED ANSWERS

Discovery
Charles Darwin announced his theory of natural selection in 1858. He published *The Origin of Species* the following year.
In 1866, Gregor Mendel, an Austrian monk, published the results of his experiments into inherited 'factors' in pea plants. He is often called the 'father of genetics'.
Walter Fleming discovered rod shaped structures in dyed cells. He called these structures 'chromosomes'. The year was 1882.
Thomas Morgan proposed the gene theory (including sex-linked inheritance) in 1910. He showed that chromosomes carry genes. Twenty-three years later, he became the first geneticist to win the Nobel Prize for Medicine.
DNA was purified by Oswald Avery, Colin MacLeod and Maclyn McCarty in 1944. They showed that DNA (not proteins) carries the genetic information in a cell.
Rosalind Franklin took X-ray diffraction photographs of DNA during 1951. Her work was crucial to the later discovery of the structure of DNA by Watson and Crick. She died three years before the scientists were awarded the Nobel Prize.
In 1953, Francis Crick and James Watson proposed the double helical structure of DNA. Nine years later, Watson and Crick, along with Maurice Wilkins, were awarded the Nobel Prize for Medicine for their discovery of the structure of DNA.
The genetic code (triplets of bases) was cracked by Marshall Nirenberg and Gobind Khorana during 1966. They showed that each amino acid is coded by a sequence of three bases (called a codon).
The Human Genome Program was launched in 1990, with the aim of sequencing all the chromosomes making up human DNA. The HGP will result in sequencing all 3.2 billion letters of the human genome.
2004 saw DNA sequencing of the rat completed. The rat has the same number of genes (25,000-30,000) as a human.

EXTENSION SUGGESTION

Copy down any new words and find their meaning.

GENES AND GENETICS
GENETICS TIMELINE

21

TASK

Read through the statements and arrange them in the correct chronological order on the timeline below.

Timeline		
	The genetic code (triplets of bases) was cracked by Marshall Nirenberg and Gobind Khorana during 1966. They showed that each amino acid is coded by a sequence of three bases (called a codon).	
	DNA was purified by Oswald Avery, Colin MacLeod and Maclyn McCarty in 1944. They showed that DNA (not proteins) carries the genetic information in a cell.	
1900 AD	In 1953, Francis Crick and James Watson proposed the double helical structure of DNA. Nine years later, Watson and Crick, along with Maurice Wilkins, were awarded the Nobel Prize for Medicine for their discovery of the structure of DNA.	
	Thomas Morgan proposed the gene theory (including sex-linked inheritance) in 1910. He showed that chromosomes carry genes. Twenty-three years later, he became the first geneticist to win the Nobel Prize for Medicine.	
1950 AD	Rosalind Franklin took X-ray diffraction photographs of DNA during 1951. Her work was crucial to the later discovery of the structure of DNA by Watson and Crick. She died three years before the scientists were awarded the Nobel Prize.	
	Walter Fleming discovered rod shaped structures in dyed cells. He called these structures 'chromosomes'. The year was 1882.	
	Charles Darwin announced his theory of natural selection in 1858. He published *The Origin of Species* the following year.	
2000 AD	2004 saw DNA sequencing of the rat completed. The rat has the same number of genes (25,000-30,000) as a human.	
	In 1866, Gregor Mendel, an Austrian monk, published the results of his experiments into inherited 'factors' in pea plants. He is often called the 'father of genetics'.	
	The Human Genome Program was launched in 1990, with the aim of sequencing all the chromosomes making up human DNA. The HGP will result in sequencing all 3.2 billion letters of the human genome.	

KS4 BIOLOGY HOW SCIENCE WORKS: COPYMASTER
Biology HSW © Badger Publishing Ltd

GENES AND GENETICS
GROWING GM COTTON IN INDIA: SOCIAL, ECONOMIC AND ENVIRONMENTAL ISSUES

22

KS4 NATIONAL CURRICULUM HSW LINK

4. *Applications and implications of science*
   b. to consider how and why decisions about science and technology are made, including those that raise ethical issues, and about the social, economic and environmental effects of such decisions

RESOURCES:
Task Sheet 22 cut into 6 cards, with instructions. Enough for one between two learners.

TIME:
10-15 minutes for the activity, 10 minutes for class discussion.

NOTES

- This activity is suitable for use as a Starter, Main activity or Plenary.
- This can be used as an introduction to the issues of GM crops or as a consolidation activity.
- Learners must be aware of these key words/concepts before attempting the task: social, economic and environmental.

SUGGESTED ANSWERS

1. "The GM cotton produces more cotton so we will have more cotton to sell and we will be able to make more money." (Ec)
2. "Scientists say they don't know the long term health effects of GM cotton – and the fields are very close to the school." (S)
3. "We have a responsibility to our customers to provide them with the best seed. The concerns about GM crops are unfounded." (S/Ec)
4. "People are concerned about the effects of GM cotton on their health and the environment – but they don't want to be left behind in the race to increase cotton yields." (S/Env)
5. "We are concerned that GM cotton could contaminate wild varieties of cotton that are native to India." (Env)
6. "The cost of GM cotton seed is over 350% higher than conventional varieties and our research shows GM cotton gives no benefit over them." (Ec)

EXTENSION SUGGESTION

Write a statement that an independent scientist might make about the growing of GM cotton in India.

Genes and Genetics
Growing GM Cotton in India: Social, Economic and Environmental Issues

22

Task

Discuss each statement card and decide whether it is a social, economic or environmental argument about GM crops.

- **Social:** To do with people's lives and the effect on running a society.
- **Economic:** To do with money, either making money or keeping costs down.
- **Environmental:** To do with keeping our environment unpolluted.

"The GM cotton produces more cotton so we will have more cotton to sell and we will be able to make more money." **Vikram, local farmer**	"Scientists say they don't know the long term health effects of GM cotton – and the fields are very close to the school." **Ushma, mother and primary school teacher**
"We have a responsibility to our customers to provide them with the best seed. The concerns about GM crops are unfounded." **Sanjay, agrochemical company representative**	"People are concerned about the effects of GM cotton on their health and the environment – but they don't want to be left behind in the race to increase cotton yields." **Vipin, government spokesperson**
"We are concerned that GM cotton could contaminate wild varieties of cotton that are native to India." **Lata, local environmental group representative**	"The cost of GM cotton seed is over 350% higher than conventional varieties and our research shows GM cotton gives no benefit over them." **Zaveri, agricultural scientist**

GENES AND GENETICS
GENETICALLY MODIFIED (GM) CROPS: BENEFITS, DRAWBACKS AND RISKS

23

KS4 NATIONAL CURRICULUM HSW LINK

4. Applications and implications of science
   a. about the use of contemporary scientific and technological developments and their benefits, drawbacks and risks

RESOURCES:
Task Sheet 23 cut into 9 cards, with instructions. Enough for one between two or four learners.

TIME:
10 minutes for the activity, 10 minutes for class discussion.

NOTES

- This activity is suitable for use as an extended Starter or Plenary or a Main activity.
- Learners must be aware of these key words/concepts before attempting the task: benefit, drawback, risk, genetic modification, GM, ecosystems, gene pool, gene.

SUGGESTED ANSWERS

Below is the intention of each statement: *Benefits: B, Drawbacks: D, Risks: R.*

1. GM can lead to higher yielding crops which means farmers can produce more from the same area of land. (B)
2. Once established, a GM crop would be almost impossible to remove from the environment. (D/R)
3. Consumers are wary of GM foods. (D)
4. GM crops or the genes themselves may escape into natural ecosystems. (R)
5. GM crops can produce healthy versions of the wild type, e.g. vegetable oils with little or no trans fatty acids – a healthier vegetable oil. (B)
6. Growing GM crops will reduce the gene pool in the wild population. (D/R)
7. Long term effects of GM crops on health are not known. (D/R)
8. The use of GM crops may prevent farmers from using plant seeds adapted to local conditions. (D)
9. GM can give farmers greater options for weed and insect control, which can lead to less soil erosion and reduced insecticide use. (B)

EXTENSION SUGGESTION

List any questions you have about the statements.

Genes and Genetics
Genetically Modified (GM) Crops: Benefits, Drawbacks and Risks

23

Task

Discuss each statement card and decide whether it is a benefit, drawback or a risk of GM crops.

- A **benefit** is something that generally has a good effect on people.
- A **drawback** is something that is a hindrance or is the 'downside'.
- A **risk** is a possible danger or source of harm.

GM can lead to higher yielding crops which means farmers can produce more from the same area of land.	Once established, a GM crop would be almost impossible to remove from the environment.	Consumers are wary of GM foods.
GM crops or the genes themselves may escape into natural ecosystems.	GM crops can produce healthy versions of the wild type, e.g. vegetable oils with little or no trans fatty acids – a healthier vegetable oil.	Growing GM crops will reduce the gene pool in the wild population.
Long term effects of GM crops on health are not known.	The use of GM crops may prevent farmers from using plant seeds adapted to local conditions.	GM can give farmers greater options for weed and insect control, which can lead to less soil erosion and reduced insecticide use.

ELECTRICAL AND CHEMICAL SIGNALS
MEASURING REACTION TIMES

24

KS4 NATIONAL CURRICULUM HSW LINK

2. *Practical and enquiry skills*
   d. *evaluate methods of collection of data and consider their validity and reliability as evidence*

RESOURCES:
Task Sheet 24, enough for one per learner or one between two learners.

TIME:
10 minutes for the activity, 10 minutes for class discussion.

NOTES

- This activity is suitable for use as a Starter or Plenary, or Main or Homework activity.
- Learners must be aware of these key words/concepts before attempting the task: precise, accurate and reliable data collection, reaction times and reflexes.

SUGGESTED ANSWERS

	Group 1	Group 2	Group 3
Accurate	Yes	No	Yes
Precise	Yes	Yes	No
Reliable	Yes	Yes	Yes

EXTENSION SUGGESTION

How could each group improve their data collection?

ELECTRICAL AND CHEMICAL SIGNALS
MEASURING REACTION TIMES

24

Three groups of students have carried out an investigation into reaction times. They used a ruler, which a student had to catch as soon as they could after it was dropped. From the distance at which the student had caught the ruler, the students calculated the reaction times. The results tables are below.

Group 1

Repeat	Reaction Time (s)
1	0.12
2	0.13
3	0.12
4	0.11
5	0.12
6	0.12
7	0.12
8	0.11
9	0.12
10	0.12

Group 2

Repeat	Reaction Time (s)
1	0.99
2	0.98
3	0.95
4	0.96
5	0.97
6	0.95
7	0.93
8	0.92
9	0.99
10	0.96

Group 3

Repeat	Reaction Time (s)
1	0.1
2	0.1
3	0.1
4	0.1
5	0.1
6	0.1
7	0.1
8	0.1
9	0.1
10	0.1

TASK

Reaction times in this activity are usually between 0.1 and 0.2 seconds. For each set of results, discuss whether you think the results are accurate, precise or reliable. Decide 'Yes' or 'No' and mark it in the table.

	Group 1	Group 2	Group 3
Accurate			
Precise			
Reliable			

ELECTRICAL AND CHEMICAL SIGNALS
DRIVER RESPONSE TIMES

25

KS4 NATIONAL CURRICULUM HSW LINK

2. *Practical and enquiry skills*
   b. collect data from primary or secondary sources, including using ICT sources and tools
   d. evaluate methods of collection of data and consider their validity and reliability as evidence

RESOURCES:
Task Sheet 25, enough for one between two learners.

TIME:
10 minutes for the activity, 5 minutes for class discussion.

NOTES

- This activity is suitable for use as a Starter or Plenary, or Main or Homework activity.
- Learners must be aware of these key words/concepts before attempting the task: response time.

SUGGESTED ANSWERS

Situation	Driver Response time (seconds)
Average driver	1.5
Drivers over 70 years old	1.8
Tired	1.8
Using a mobile phone	1.8-2.5
Drunk driver	3.0

EXTENSION SUGGESTION

Draw a graph of the results.

ELECTRICAL AND CHEMICAL SIGNALS
DRIVER RESPONSE TIMES

25

TASK

Read the text below and put the information into a table that shows drivers' response times in different situations.

> The average driver in normal conditions has a response time of 1.5 seconds. That is the time it takes to notice the hazard, decide to react and push the break pedal. There is no significant difference between male and female response times, however drivers over 70 years old do have an increased response time of 1.8 seconds. This delay is the same for drivers who are tired.
>
> Drivers who are distracted, for example using a mobile phone, take an additional 0.3 to 1 second to respond. Drunk drivers can take up to 3 seconds to respond to a hazard.
>
> Even two drivers with the same response time can have different overall stopping distances if they are travelling at different speeds. A car travelling at 65km per hour will travel 27.1 metres in 1.5 seconds, however at 5km/h slower, the same driver would travel only 25m in the 1.5 seconds response time.

ELECTRICAL AND CHEMICAL SIGNALS
TIRED REACTIONS

26

KS4 NATIONAL CURRICULUM HSW LINK

2. Practical and enquiry skills
   a. plan to test a scientific idea, answer a scientific question or solve a scientific problem
   b. work accurately and safely, individually and with others, when collecting first-hand data

RESOURCES:
Task Sheet 26, enough for one between two learners.

TIME:
10 minutes for the activity, 10 minutes for class discussion.

NOTES

- This activity is suitable for use as a Starter or Plenary, or Main or Homework activity.
- Learners must be aware of these key words/concepts before attempting the task: valid, accurate and reliable data collection, reaction times and reflexes.

SUGGESTED ANSWERS

A. Prediction with a scientific reason.
   The more tired someone is, the slower their reaction time.
B. Bullet point method (how you will do the experiment):
   1) *Use the same people.*
   2) *Measure reactions at specific times of the day.*
   3) *Make sure people have not eaten or drunk substances that could affect their reaction times, e.g. containing caffeine.*
C. State how you will ensure that the experiment is:
   - valid – *use same people; check dietary intake before testing*;
   - reliable – *repeat readings*;
   - accurate – *computer timing is accurate*
D. Safety considerations:
   Proximity to computer screen.

EXTENSION SUGGESTION

Draw a table to record the results in.

ELECTRICAL AND CHEMICAL SIGNALS
TIRED REACTIONS

26

Scientists use some computer software that can measure reaction times.

The person the scientists are testing sits in front of a blank computer screen, with one finger over the space bar. When a red square flashes up on the screen, the person has to press the space bar as quickly as they can.

The computer measures the time between the red square appearing and the person pressing the space bar.

This is repeated ten times and an average reaction time is calculated.

TASK

Plan an investigation to find out the effect of tiredness on reaction times.

Include:

A. Prediction with a scientific reason.
B. Bullet point method (how you will do the experiment).
C. State how you will ensure that the experiment is:
- valid
- reliable
- accurate
D. Safety considerations.

ELECTRICAL AND CHEMICAL SIGNALS
IMPROVING REACTION TIMES

27

KS4 NATIONAL CURRICULUM HSW LINK

3. Communication skills
    a. recall, analyse, interpret, apply and question scientific information or ideas

RESOURCES:
Task Sheet 27, enough for one between two learners.

TIME:
10 minutes for the activity, 10 minutes for class discussion.

NOTES

- This activity is suitable for use as a Starter or Plenary, or Main or Homework activity.
- Learners must be aware of these key words/concepts before attempting the task: reaction times, effects of caffeine.

SUGGESTED ANSWERS

A. Get plenty of the sleep before the race.
   Useful advice because tiredness slows reactions.
B. Eat plenty of carbohydrate before the race.
   Not useful.
C. Drink plenty of drinks containing caffeine.
   Not useful.
D. Practice regularly at responding to the starting pistol.
   Useful as you can improve reactions through practice and anticipation.
E. Warm up thoroughly before the race.
   Useful as increased blood flow and adrenaline will increase response times.

EXTENSION SUGGESTION

Suggest any further advice to give to a sprinter.

ELECTRICAL AND CHEMICAL SIGNALS
IMPROVING REACTION TIMES

27

Many sports people train hard to improve their reaction time. This is particularly important for short-distance runners, when every millisecond counts. Other sports where reaction time is important include football, baseball and cricket.

TASK

Read each of these pieces of advice. Decide if each one would help improve a sprinter's reaction time to the starting pistol. Give a scientific reason for each answer.

A. Get plenty of the sleep before the race.

B. Eat plenty of carbohydrate before the race.

C. Drink plenty of drinks containing caffeine.

D. Practice regularly at responding to the starting pistol.

E. Warm up thoroughly before the race.

ELECTRICAL AND CHEMICAL SIGNALS
BLOOD SUGAR GRAPHS

28

KS4 NATIONAL CURRICULUM HSW LINK

3. Communication skills
   a. recall, analyse, interpret, apply and question scientific information or ideas

RESOURCES:
Task Sheet 28, enough for one per learner or one between two learners.

TIME:
10 minutes for the activity, 5 minutes for class discussion.

NOTES

- This activity is suitable for use as a Starter or Plenary, or Main or Homework activity.
- Learners must be aware of these key words/concepts before attempting the task: blood sugar level, diabetes, insulin.

SUGGESTED ANSWERS

Graph 1: C
Graph 2: B
Graph 3: A
Graph 4: D

EXTENSION SUGGESTION

Explain how insulin controls blood sugar levels.

KS4 BIOLOGY HOW SCIENCE WORKS: TEACHER NOTES

ELECTRICAL AND CHEMICAL SIGNALS
BLOOD SUGAR GRAPHS

28

TASK

Match the graphs to the correct titles.

A. The blood sugar level of a person using insulin to control diabetes.
B. The blood sugar level of a person with a healthy pancreas.
C. The blood sugar level of a person becoming hyperglycaemic.
D. The blood sugar level of a person controlling diabetes through diet.

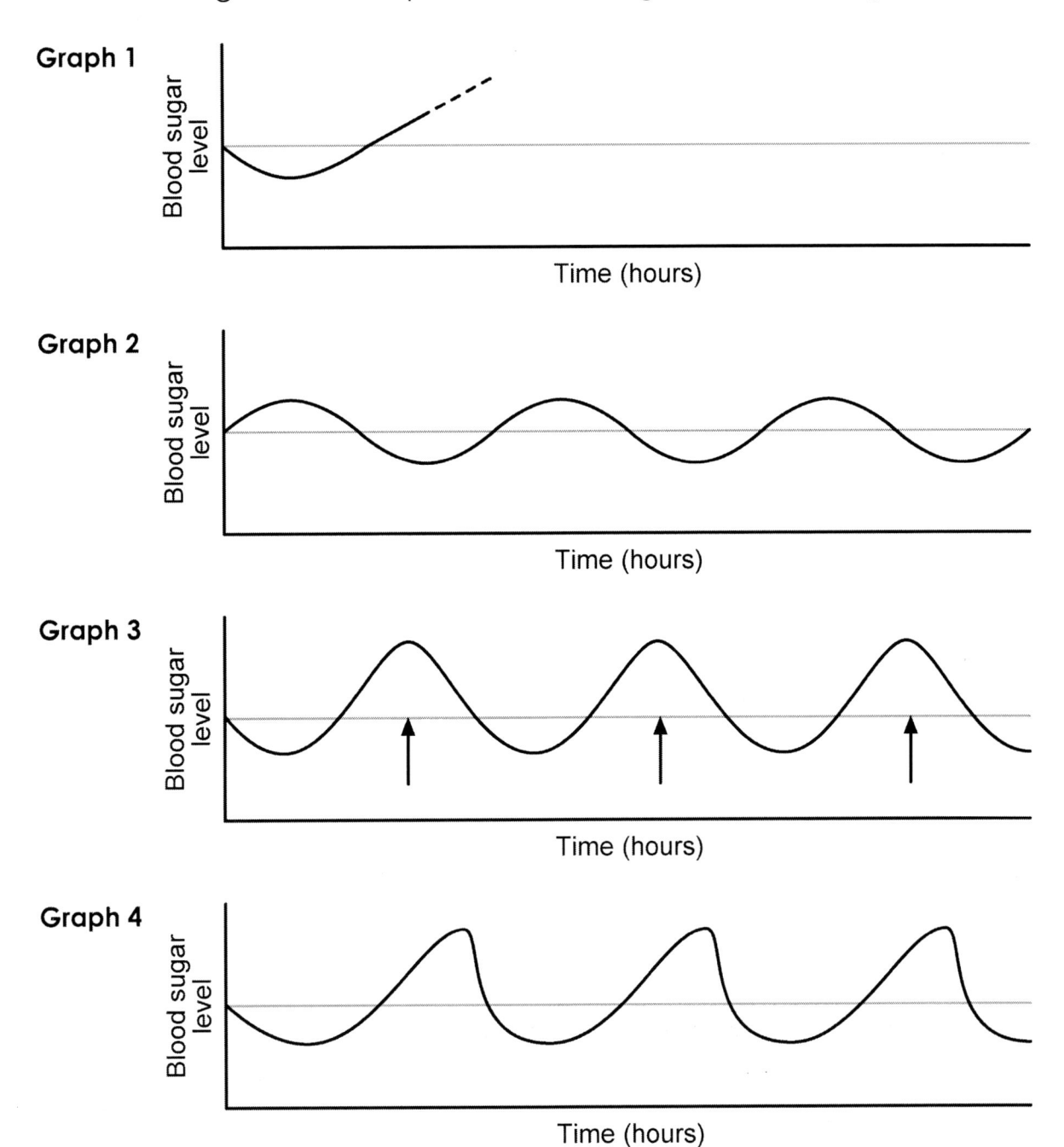

Graph 1

Graph 2

Graph 3

Graph 4

ELECTRICAL AND CHEMICAL SIGNALS
FEMALE HORMONE GRAPH ANALYSIS
29

KS4 NATIONAL CURRICULUM HSW LINK

3. Communication skills
   a. recall, analyse, interpret, apply and question scientific information or ideas
   b. use both qualitative and quantitative approaches
   c. present information, develop an argument and draw a conclusion, using scientific, technical and mathematical language, conventions and symbols, and ICT tools

RESOURCES:
Task Sheet 29, enough for one between two learners.

TIME:
10 minutes for the activity, 10 minutes for class discussion.

NOTES

- This activity is suitable as a Main or Homework activity, or Plenary. This task could also be used as a stimulus activity when introducing learners to the topic of hormones and their use in controlling female fertility.
- Learners must be aware of these key words/concepts before attempting the task: the structure of the female reproductive organs, hormone, fertility, FSH, LH, maturation, ovulation, uterus, ovary.

SUGGESTED ANSWERS

A. Which data line corresponds to which hormone?
   Make sure you can explain your decisions.

   Discuss learner's reasons for their decision in relation to the text below the graph.

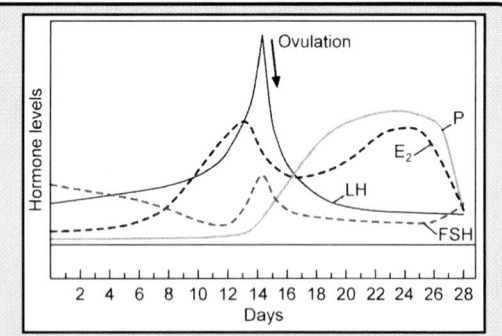

B. Can you suggest why high levels of FSH are given to women undergoing IVF treatment?
   To stimulate the maturation of many eggs.

C. The contraceptive pill contains synthetic oestrogen. How will this additional oestrogen affect a woman's fertility?
   High levels of oestrogen inhibit the production of FSH so mature eggs will not develop.

EXTENSION SUGGESTION

Copy down any new words and find their meaning.

ELECTRICAL AND CHEMICAL SIGNALS
FEMALE HORMONE GRAPH ANALYSIS
29

TASK

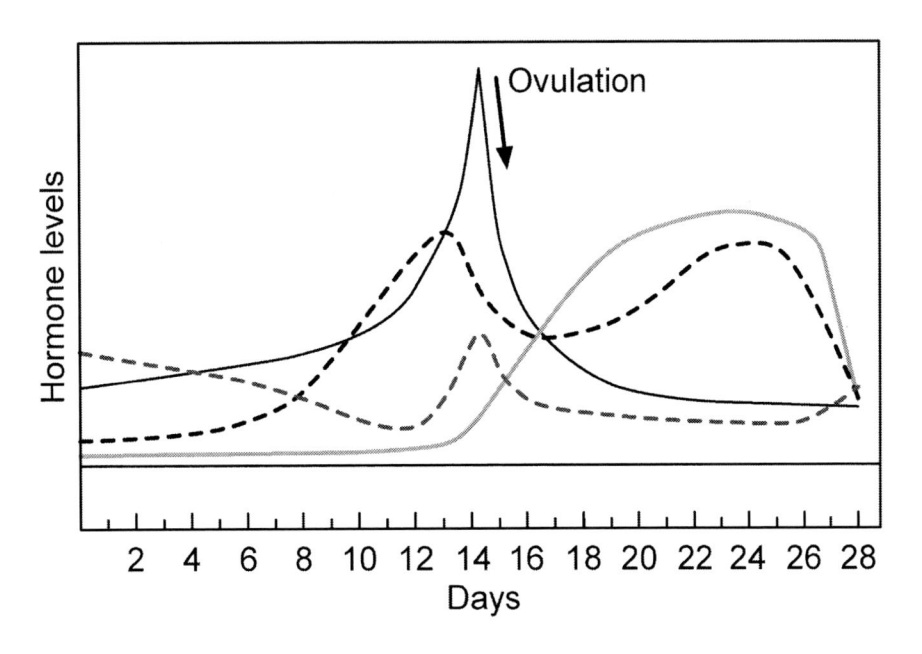

Key
- *Follicle Stimulating Hormone (FSH)* – FSH stimulates an ovary to mature one egg during the first two weeks of the cycle. It stimulates the hormone oestrogen to be secreted by the maturing egg. Levels of FSH drop slowly and then peak at ovulation.
- *Luteinising Hormone (LH)* – this hormone is responsible for the release of the mature egg from the ovary. Its production is stimulated by oestrogen.
- *Oestrogen* – this hormone encourages the lining of the uterus to thicken before ovulation and prevents any more eggs from developing. Its secretion from the ovary is stimulated by high levels of FSH.
- *Progesterone* – progesterone levels increase rapidly after ovulation to maintain the thick lining of the uterus in case the egg is fertilised. Levels of progesterone remain high until the unfertilised egg dies, whereupon levels decrease rapidly and the lining of the uterus breaks down.

Using the graph and key, answer these questions:

A. Which data line corresponds to which hormone? Make sure you can explain your decisions.
B. Can you suggest why high levels of FSH are given to women undergoing IVF treatment?
C. The contraceptive pill contains synthetic oestrogen. How will this additional oestrogen affect a woman's fertility?

ELECTRICAL AND CHEMICAL SIGNALS
DIABETES TIMELINE

30

KS4 NATIONAL CURRICULUM HSW LINK

4. Applications and implications of science
   c. how uncertainties in scientific knowledge and scientific ideas change over time and about the role of the scientific community in validating these changes

RESOURCES:
Task Sheet 30, cut into 16 cards, with instructions. Enough for one per learner.

TIME:
15 minutes for the activity, 5 minutes for class discussion.

NOTES

- This activity is suitable for use as a Main or Homework activity.
- Learners must be aware of these key words/concepts before attempting the task: diabetes, role of insulin in controlling blood sugar levels.

SUGGESTED ANSWERS

As early as 1550 BC, Ancient Egyptian papyrus mentioned a rare disease that caused patients to urinate frequently and lose weight rapidly.
In 1 AD, Greek physician Aretaeus recorded a disease with symptoms of frequent urination, constant thirst and weight loss. He called the condition 'diabetes', meaning 'siphon' or 'a flowing through'.
During the 17th century, urine was used to diagnose various disorders, including diabetes. Some physicians, like Dr. Thomas Willis, tasted the urine and detected a sweet taste, giving the disease a second name, 'mellitis', meaning honey. This method of testing the urine carried on until the 20th century.
During the Franco-Prussian war (1870s), Bouchardat, a French physician, discovered a link between diet (calorie intake) and diabetes. He found that a reduced diet helped his patients.
In 1869, Paul Langerhans, a medical student, examined the structure of the pancreas and identified tiny structures of cells with an unknown function.
Moses Barron, an American, linked the Langerhans cells in the pancreas with diabetes in 1920.
One year later, in 1921, Canadian Frederick Banting carried out a series of experiments to link diabetes with the pancreas. He discovered the role of the hormone 'insulin' by using it to treat a de-pancreatized dog. In 1923, Banting was awarded the Nobel Prize in Physiology and Medicine.
In 1922, a 14 year old boy called Leonard Thompson was the first human treated with beef insulin extracts. The treatment was declared a success. Prior to the discovery of insulin, a child diagnosed with diabetes was normally given a life expectancy of less than one year.
During 1935, Roger Hinsworth discovered there are two forms of diabetes: type I (insulin sensitive) and type II (insulin insensitive). This opened new avenues for treatment.
1983 saw human insulin produced by genetically modified bacteria approved by the American Food and Drug Administration (FDA). It replaced insulin extracted from beef and pork sources.

EXTENSION SUGGESTION

Copy down any new words and find their meaning.

KS4 BIOLOGY HOW SCIENCE WORKS: TEACHER NOTES

ELECTRICAL AND CHEMICAL SIGNALS
DIABETES TIMELINE

30

TASK

Read through the statements and arrange them in the correct chronological order on the timeline below.

Timeline		
	During the 17th century, urine was used to diagnose various disorders, including diabetes. Some physicians, like Dr. Thomas Willis, tasted the urine and detected a sweet taste, giving the disease a second name, 'mellitis', meaning honey. This method of testing the urine carried on until the 20th century.	
	During 1935, Roger Hinsworth discovered there are two forms of diabetes: type I (insulin sensitive) and type II (insulin insensitive). This opened new avenues for treatment.	
— 1700 AD	In 1869, Paul Langerhans, a medical student, examined the structure of the pancreas and identified tiny structures of cells with an unknown function.	
	During the Franco-Prussian war (1870s), Bouchardat, a French physician, discovered a link between diet (calorie intake) and diabetes. He found that a reduced diet helped his patients.	
— 1800 AD	One year later, in 1921, Canadian Frederick Banting carried out a series of experiments to link diabetes with the pancreas. He discovered the role of the hormone 'insulin' by using it to treat a de-pancreatized dog. In 1923, Banting was awarded the Nobel Prize in Physiology and Medicine.	
	Moses Barron, an American, linked the Langerhans cells in the pancreas with diabetes in 1920.	
— 1900 AD		
	As early as 1550 BC, Ancient Egyptian papyrus mentioned a rare disease that caused patients to urinate frequently and lose weight rapidly.	
	In 1 AD, Greek physician Aretaeus recorded a disease with symptoms of frequent urination, constant thirst and weight loss. He called the condition 'diabetes', meaning 'siphon' or 'a flowing through'.	
— 2000 AD		
	1983 saw human insulin produced by genetically modified bacteria approved by the American Food and Drug Administration (FDA). It replaced insulin extracted from beef and pork sources.	
	1922, a 14 year old boy called Leonard Thompson was the first human to be treated with beef insulin extracts. The treatment was a success. Prior to the discovery of insulin, a child diagnosed with diabetes was normally given a life expectancy of less than one year.	

KS4 BIOLOGY HOW SCIENCE WORKS: COPYMASTER

Biology HSW © Badger Publishing Ltd

ELECTRICAL AND CHEMICAL SIGNALS
THE CONTRACEPTIVE PILL: SOCIAL, ECONOMIC AND ENVIRONMENTAL ISSUES

31

KS4 NATIONAL CURRICULUM HSW LINK

4. *Applications and implications of science*
    b. to consider how and why decisions about science and technology are made, including those that raise ethical issues, and about the social, economic and environmental effects of such decisions

RESOURCES:
Task Sheet 31 cut into 6 cards, with instructions. Enough for one between two learners.

TIME:
10-15 minutes for the activity, 10 minutes for class discussion.

NOTES

- This activity is suitable for use as a Starter, Main activity or Plenary.
- Learners must be aware of these key words/concepts before attempting the task: social, economic and environmental.

SUGGESTED ANSWERS

1. The contraceptive pill has changed women's lives by giving them control over their own fertility. **(S)**
2. I think the contraceptive pill is wrong morally – it goes against the teachings of my religion. **(S)**
3. The ever increasing human population puts a tremendous strain on the environment and its natural resources. **(Ev)**
4. The contraceptive pill reduces the number of unwanted pregnancies and babies which would otherwise be a burden on the welfare state. **(Ec)**
5. The contraceptive pill allowed me to choose when to have children but I am not happy about it being prescribed to my daughter without my knowledge. **(S/Ec)**
6. The contraceptive pill has increased the probability of a woman working, the investment in her education, income levels and satisfaction with life. **(S/Ec/Ev)**

EXTENSION SUGGESTION

Write a statement to show how you think the contraceptive pill has changed women's lives over the past 50 years.

ELECTRICAL AND CHEMICAL SIGNALS
THE CONTRACEPTIVE PILL: SOCIAL, ECONOMIC AND ENVIRONMENTAL ISSUES

31

TASK

Discuss each statement card and decide whether it is a social, economic or environmental issue about the contraceptive pill.

- **Social:** To do with people's lives and the effect on running a society.
- **Economic:** To do with money, either making money or keeping costs down.
- **Environmental:** To do with keeping our environment unpolluted.

"The contraceptive pill has changed women's lives by giving them control over their own fertility." **Sue, agony aunt**	"I think the contraceptive pill is wrong morally – it goes against the teachings of my religion." **Jonathan, religious group leader**
"The ever increasing human population puts a tremendous strain on the environment and its natural resources." **Joanne, environmental campaigner**	"The contraceptive pill reduces the number of unwanted pregnancies and babies which would otherwise be a burden on the welfare state." **Mike, politician**
"The contraceptive pill allowed me to choose when to have children but I am not happy about it being prescribed to my daughter without my knowledge." **Sandra, mother of a teenage daughter**	"The contraceptive pill has increased the probability of a woman working, the investment in her education, income levels and satisfaction with life." **Ushma, professor of women's education**

ELECTRICAL AND CHEMICAL SIGNALS
In Vitro Fertilisation (IVF): Benefits, Drawbacks and Risks

32

KS4 National Curriculum HSW link

4. Applications and implications of science
   a. about the use of contemporary scientific and technological developments and their benefits, drawbacks and risks

RESOURCES:
Task Sheet 32 cut into 9 cards, with instructions. Enough for one between two or four learners.

TIME:
10 minutes for the activity, 10 minutes for class discussion.

Notes

- This activity is suitable for use as an extended Starter or Plenary, or Main activity.
- Learners must be aware of these key words/concepts before attempting the task: benefit, drawback and risk, IVF, infertility, multiple births.

Suggested answers

Below is the intention of each statement: *Benefits: B, Drawbacks: D, Risks: R.*

1. IVF has allowed people made infertile through cancer treatment to have children. *(B)*
2. IVF may increase the chance of a genetic defect being passed to a child. *(R)*
3. Difficult decisions may have to be made if excess embryos are produced during a cycle of IVF. *(D)*
4. IVF often results in multiple births. *(B/D/R)*
5. IVF requires the woman to take large amounts of drugs to stimulate her ovaries to mature lots of eggs. *(D/R)*
6. IVF enables infertile people to have children. *(B)*
7. The success rate of IVF is only about 25%. *(D)*
8. IVF may increase the chance of a woman developing certain types of cancer. *(R)*
9. IVF is very expensive. *(D)*

Extension suggestion

List any questions you have about the statements.

Electrical and Chemical Signals
In Vitro Fertilisation (IVF): Benefits, Drawbacks and Risks

32

Task

Discuss each statement card and decide whether it is a benefit, drawback or a risk of IVF treatment.

- A **benefit** is something that generally has a good effect on people.
- A **drawback** is something that is a hindrance or is the 'downside'.
- A **risk** is a possible danger or source of harm.

IVF has allowed people made infertile through cancer treatment to have children.	IVF may increase the chance of a genetic defect being passed on to a child.	Difficult decisions may have to be made if excess embryos are produced during a cycle of IVF.
IVF often results in multiple births.	IVF requires the woman to take large amounts of drugs to stimulate her ovaries to mature lots of eggs.	IVF enables infertile people to have children.
The success rate of IVF is only about 25%.	IVF may increase the chance of a woman developing certain types of cancer.	IVF is very expensive.

KS4 Biology How Science Works: Copymaster

Biology HSW © Badger Publishing Ltd

# HEALTH, MEDICINE AND DRUGS
CANNABIS & TOBACCO – THE EVIDENCE

33

KS4 NATIONAL CURRICULUM HSW LINK

1. Data, evidence, theories and explanations
   a. how scientific data can be collected and analysed
   b. how interpretation of data, using creative thought, provides evidence to test ideas and develop theories
   c. how explanations of many phenomena can be developed using scientific theories, models and ideas
   d. that there are some questions that science cannot currently answer, and some that science cannot address

RESOURCES:
Task Sheet 33 cut into 9 cards, with instructions. Enough for one between two or four learners.

TIME:
10 minutes for the activity, 10 minutes for class discussion.

NOTES

- This activity is suitable for use as a Starter, Main activity or Plenary.
- Learners must be aware of these key words/concepts before attempting the task: evaluation, evidence, multiple sclerosis, hallucinations and cause paranoia, schizophrenia.

SUGGESTED ANSWERS

1. Some people claim that smoking cannabis helps relieve symptoms of diseases such as multiple sclerosis. *(None)*
2. In addition to the cancer causing chemicals in tobacco smoke, cannabis contains chemicals that can slow reactions, cause hallucinations and cause paranoia. *(For)*
3. Regular use of cannabis can lead to general tiredness and lack of motivation to do anything. *(None/For)*
4. Smoking cannabis in the UK is illegal, whereas smoking tobacco is not. *(None)*
5. There is increasing evidence that cannabis can lead to mental health conditions such as schizophrenia in some people. *(For)*
6. Some cannabis users smoke stronger varieties of cannabis to get a bigger 'high'; the stronger the cannabis, the more risk there is of mental health effects. *(For)*
7. Cannabis can have a similar effect on a user to drinking alcohol, it slows reactions, making driving dangerous. It is safe to drive after smoking tobacco. *(For)*
8. Regular use of cannabis can reduce sperm counts and delay ovulation. Regular tobacco smokers can suffer from reduced fertility. *(For)*
9. Nicotine in tobacco is addictive. There are no addictive chemicals in cannabis, but users can become dependent on smoking cannabis. *(None)*

EXTENSION SUGGESTION

Suggest the evidence that people would need to show that cannabis helps relieve the symptoms of some diseases, such as multiple sclerosis.

KS4 BIOLOGY HOW SCIENCE WORKS: Teacher notes

HEALTH, MEDICINE AND DRUGS
CANNABIS & TOBACCO – THE EVIDENCE

33

Some doctors have a theory that:

Smoking cannabis is more harmful to health than smoking tobacco.

TASK

Using the cards below, sort the statements into three groups:
- Evidence for the theory.
- Evidence against the theory.
- No evidence for the theory.

Some people claim that smoking cannabis helps relieve symptoms of diseases such as multiple sclerosis.	In addition to the cancer causing chemicals in tobacco smoke, cannabis contains chemicals that can slow reactions, cause hallucinations and cause paranoia.	Regular use of cannabis can lead to general tiredness and lack of motivation to do anything.
Smoking cannabis in the UK is illegal, whereas smoking tobacco is not.	There is increasing evidence that cannabis can lead to mental health conditions such as schizophrenia in some people.	Some cannabis users smoke stronger varieties of cannabis to get a bigger 'high'; the stronger the cannabis, the more risk there is of mental health effects.
Cannabis can have a similar effect on a user to drinking alcohol, it slows reactions, making driving dangerous. It is safe to drive after smoking tobacco.	Regular use of cannabis can reduce sperm counts and delay ovulation. Regular tobacco smokers can suffer from reduced fertility.	Nicotine in tobacco is addictive. There are no addictive chemicals in cannabis, but users can become dependent on smoking cannabis.

KS4 BIOLOGY HOW SCIENCE WORKS: COPYMASTER

Biology HSW © Badger Publishing Ltd

HEALTH, MEDICINE AND DRUGS
DRUG TRIALS

34

KS4 NATIONAL CURRICULUM HSW LINK

1. Data, evidence, theories and explanations
   a. how scientific data can be collected and analysed
   b. how interpretation of data, using creative thought, provides evidence to test ideas and develop theories
   c. how explanations of many phenomena can be developed using scientific theories, models and ideas

RESOURCES:
Task Sheet 34 cut into 10 cards, with instructions. Enough for one between two or four learners.

TIME:
10 minutes for the activity, 10 minutes for class discussion.

NOTES

- This activity is suitable for use as an extended Starter or Plenary, or Main or Homework activity.
- Learners must be aware of these key words/concepts before attempting the task: in vitro.

SUGGESTED ANSWERS

1. A potential new drug is identified.
2. Variations of the potential drug are made in the laboratory, often using computers to design the active ingredient.
3. The potential drugs are then tested on cell cultures, tissue cultures and isolated organs. This is called *in vitro* screening.
4. Test the potential drugs on animals to see short-term effects.
5. Human trials – test on a few healthy patients.
6. Human trials – test on a few patients with the disease.
7. Human trials – test on a large number of patients.
8. Test the potential drugs on animals to see long-term effects.*
9. If the drug has passed these tests, it will be granted a license as a new drug for doctors to use.
10. The effect of the licensed drug is monitored on all patients that use it.

* These may start at the same time or directly after animal testing for short-term effects.

EXTENSION SUGGESTION

Suggest why developing a drug costs a lot of money.

HEALTH, MEDICINE AND DRUGS
DRUG TRIALS

34

TASK

The development of new drugs is a long and risky process. Many chemicals are very harmful to cells and organs and can have nasty side effects.

Read each statement carefully, then put them in order of how you think scientists develop and test new drugs.

Human trials – test on a few patients with the disease.
Variations of the potential drug are made in the laboratory, often using computers to design the active ingredient.
A potential new drug is identified.
Human trials – test on a few healthy patients.
If the drug has passed these tests, it will be granted a license as a new drug for doctors to use.
Test the potential drugs on animals to see long-term effects.
Human trials – test on a large number of patients.
The effect of the licensed drug is monitored on all patients that use it.
The potential drugs are then tested on cell cultures, tissue cultures and isolated organs. This is called *in vitro* screening.
Test the potential drugs on animals to see short-term effects.

HEALTH, MEDICINE AND DRUGS
SMOKING STATISTICS

35

KS4 NATIONAL CURRICULUM HSW LINK

3. Communication skills
   a. recall, analyse, interpret, apply and question scientific information or ideas
   b. use both qualitative and quantitative approaches
   c. present information, develop an argument and draw a conclusion, using scientific, technical and mathematical language, conventions and symbols, and ICT tools

RESOURCES:
Task Sheet 35, enough for one per learner or one between two learners.

TIME:
10 minutes for the activity, 5 minutes for class discussion.

NOTES

- This activity is suitable for use as a Starter or Plenary, or Main or Homework activity.
- Learners must be aware of these key words/concepts before attempting the task: analysing data.

SUGGESTED ANSWERS

A. Identify the dependent and independent variables.
   Dependent = Number of deaths (due to heart attack), Independent = Year

B. Discuss whether this is the best way to present this information. How could it be improved?
   Data from previous years would help to see the trend.

C. Does this information provide sufficient evidence for the smoking ban reducing the number of deaths from heart attacks?
   There may be other factors. It shows a correlation, not a cause and effect.

D. Calculate the percentage decrease in the number of deaths from heart attacks between 2004 and 2005.
   11%

EXTENSION SUGGESTION

Suggest other factors that could contribute to the reduction in deaths due to heart attacks.

HEALTH, MEDICINE AND DRUGS
SMOKING STATISTICS

35

In Italy, 2004, there was a ban on smoking in public places.

Doctors had found that the number of deaths due to heart attacks was increasing. However, after the ban, the number of deaths due to heart attacks decreased (Figure 1).

Figure 1. Number of deaths due to heart attacks in Italy in 2004 and 2005

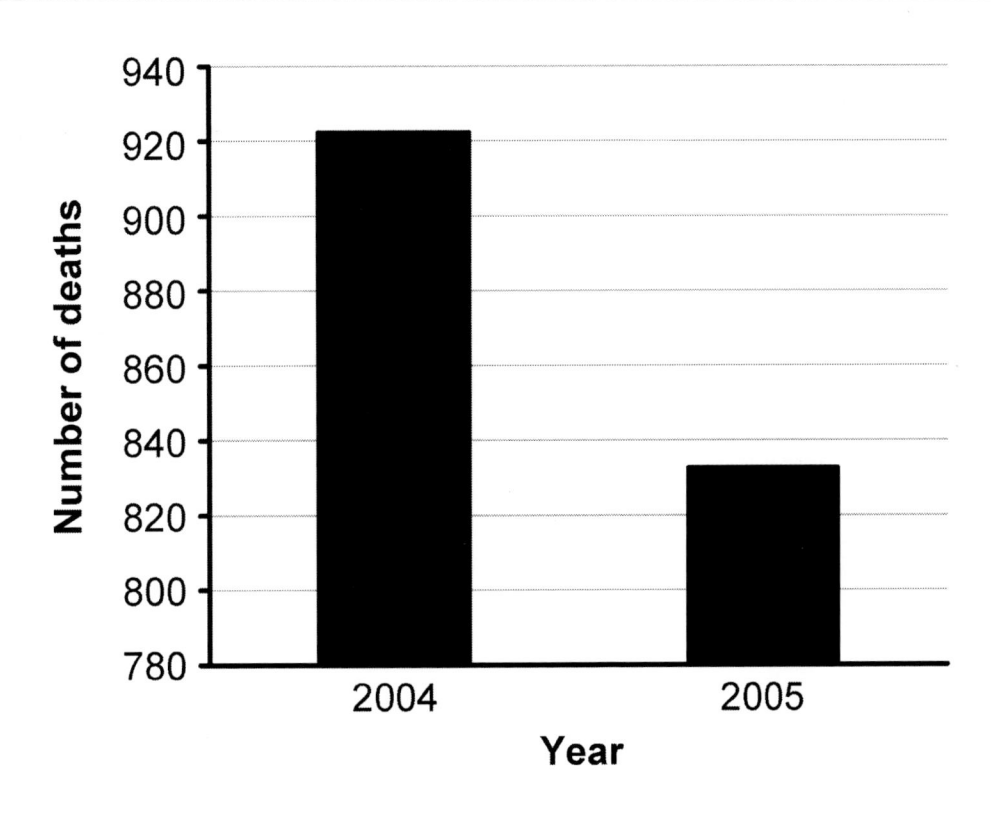

[Bar chart showing approximately 923 deaths in 2004 and approximately 833 deaths in 2005. Y-axis: Number of deaths (780–940). X-axis: Year.]

TASK

A. Identify the dependent and independent variables.
B. Discuss whether this is the best way to present this information. How could it be improved?
C. Does this information provide sufficient evidence for the smoking ban reducing the number of deaths from heart attacks?
D. Calculate the percentage decrease in the number of deaths from heart attacks between 2004 and 2005.

HEALTH, MEDICINE AND DRUGS
ANTIBIOTIC RESISTANCE

36

KS4 NATIONAL CURRICULUM HSW LINK

3. Communication skills
   a. recall, analyse, interpret, apply and question scientific information or ideas
   b. use both qualitative and quantitative approaches
   c. present information, develop an argument and draw a conclusion, using scientific, technical and mathematical language, conventions and symbols, and ICT tools

RESOURCES:
Task Sheet 36, enough for one per learner or one between two. Graph paper, rulers.

TIME:
10-15 minutes for the activity, 5 minutes for class discussion.

NOTES

- This activity is suitable for use as a Starter or Plenary, or Main or Homework activity.
- Learners must be aware of these key words/concepts before attempting the task: graph drawing, graph interpretation.

SUGGESTED ANSWERS

A. Decide how to present the data. Carefully draw your graph.
   Learners should draw a line graph, but some may choose to draw a histogram.
B. Describe the trend in the data from your graph.
   The number of deaths caused by MRSA has increased between 1993 and 2006 by about four times.
C. Does the theory above explain the trend?
   It is one possible explanation, but does not tell you that more bacteria are resistant to antibiotics.
D. What other reasons could there be for this trend?
   Increased reporting and awareness of MRSA.
E. Explain the data using your knowledge and understanding of microbes, health and data.
   Discussion about the idea of bacteria becoming resistant and then the resistant ones multiplying. An awareness of more antibiotics are used in hospitals, so it is more likely that the bacteria will become resistant to them.

EXTENSION SUGGESTION

Suggest ways to reduce the number of cases of MRSA.

HEALTH, MEDICINE AND DRUGS
ANTIBIOTIC RESISTANCE

36

MRSA is a bacterium that has shown increased resistance to antibiotics in hospitals. As more bacteria become resistant, more people will die from infection because antibiotics will be useless in treating it.

This is one explanation for the data below (Table 1).

Table 1: The number of deaths due to MRSA infection between 1993-2006

Year	Number of deaths caused by MRSA
1993	450
1994	480
1995	532
1996	621
1997	767
1998	879
1999	998
2000	1190
2001	1237
2002	1242
2003	1403
2004	1678
2005	2153
2006	2162

TASK

A. Decide how to present the data. Carefully draw your graph.
B. Describe the trend in the data from your graph.
C. Does the theory above explain the trend?
D. What other reasons could there be for this trend?
E. Explain the data using your knowledge and understanding of microbes, health and data.

HEALTH, MEDICINE AND DRUGS
TREATING TB

37

KS4 NATIONAL CURRICULUM HSW LINK

3. Communication skills
    a. recall, analyse, interpret, apply and question scientific information or ideas
    b. use both qualitative and quantitative approaches
    c. present information, develop an argument and draw a conclusion, using scientific, technical and mathematical language, conventions and symbols, and ICT tools

RESOURCES:
Task Sheet 37, enough for one per learner or one between two.

TIME:
5-10 minutes for the activity, 5 minutes for class discussion.

NOTES

- This activity is suitable for use as a Starter or Plenary, or Main or Homework activity.
- Learners must be aware of these key words/concepts before attempting the task: graph interpretation, notification (cases of TB reported).

SUGGESTED ANSWERS

A. Identify the dependent and independent variables.
   Dependent = Number (of deaths/notifications), Independent = Year
B. Suggest when antibiotic treatment became common for TB patients.
   Late 1940s – early 1950s (1955 is when deaths stayed at their minimum.)
   Actual date of use of antibiotics to treat critical TB was 1944.
C. Suggest when the BCG vaccination against TB was given to the population.
   The vaccination programme started in 1953. From the graph, the best guess of the notifications decreasing is when the numbers consistently decrease between 1952 and 1955.

EXTENSION SUGGESTION

Suggest how the graph will change as more people travel to countries where TB is very common.

KS4 BIOLOGY HOW SCIENCE WORKS: Teacher notes

Health, Medicine and Drugs
Treating TB

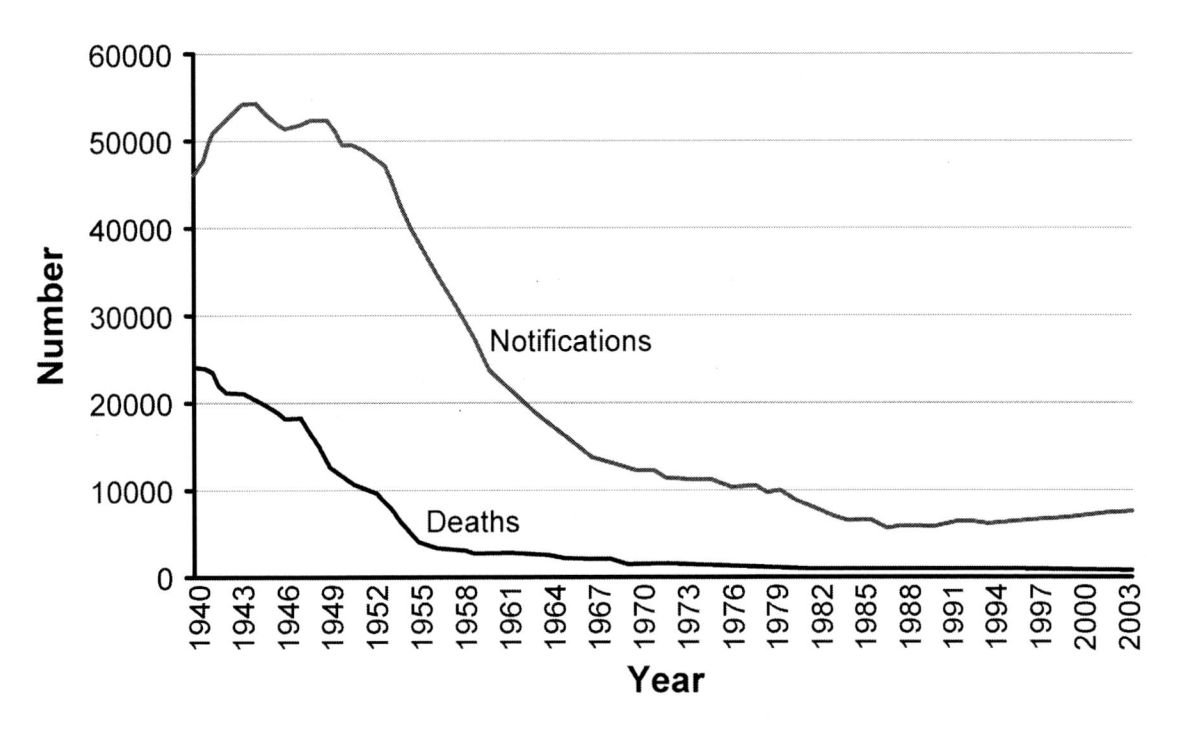

37

TB notifications and deaths in England and Wales, 1940-2003

Task

Look at the graph.

A. Identify the dependent and independent variables.
B. Suggest when antibiotic treatment became common for TB patients.
C. Suggest when the BCG vaccination against TB was given to the population.

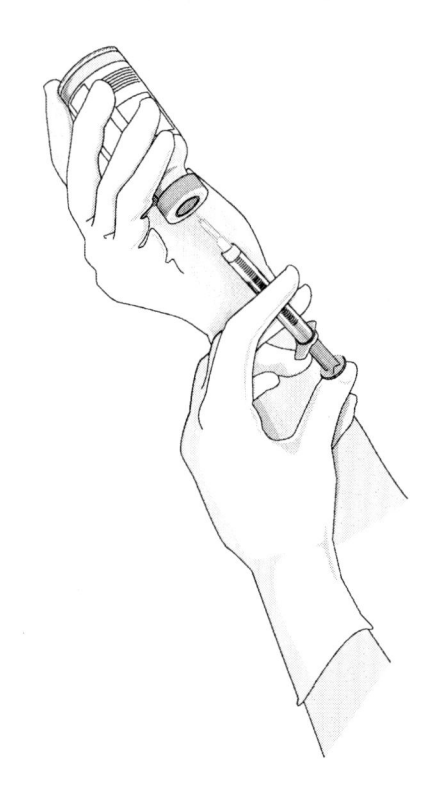

HEALTH, MEDICINE AND DRUGS
WEIGHTS & MEASURES IN MEDICINE

38

KS4 NATIONAL CURRICULUM HSW LINK

3. *Communication skills*
   b. use both qualitative and quantitative approaches
   c. present information, develop an argument and draw a conclusion, using scientific, technical and mathematical language, conventions and symbols, and ICT tools

RESOURCES:
Task Sheet 38 cut into 18 cards, with instructions. Enough for one between two learners.

TIME:
5 minutes for the activity, 5 minutes for class discussion.

NOTES

- This activity is suitable for use as a Starter, Main activity or Plenary.
- Learners must be aware of these key words/concepts before attempting the task: measurements of length, volume and mass.

SUGGESTED ANSWERS

m	metres	height of a person
cm^3	cubic centimetres	volume of solution
μm	micrometres	length of a cell
mg	milligrams	mass of tablets prescribed of drugs
kg	kilograms	mass of a person
mg per $100cm^3$	milligrams per 100 cubic centimetres	mass of alcohol per volume of blood

EXTENSION SUGGESTION

Sort into mass and length measurements, then sort into order of increasing size.

KS4 BIOLOGY HOW SCIENCE WORKS: Teacher notes

HEALTH, MEDICINE AND DRUGS
WEIGHTS & MEASURES IN MEDICINE

38

TASK

Match the cards: unit symbol, unit name and measurement.

kg	cubic centimetres	mass of tablets prescribed of drugs
mg per 100cm^3	kilograms	volume of solution
m	milligrams per 100 cubic centimeters	height of a person
cm^3	micrometres	mass of a person
μm	milligrams	length of a cell
mg	metres	mass of alcohol per volume of blood

KS4 BIOLOGY HOW SCIENCE WORKS: COPYMASTER

Biology HSW © Badger Publishing Ltd

HEALTH, MEDICINE AND DRUGS
TB TIMELINE

39

KS4 NATIONAL CURRICULUM HSW LINK

4. Applications and implications of science
   c. how uncertainties in scientific knowledge and scientific ideas change over time and about the role of the scientific community in validating these changes

RESOURCES:
Task Sheet 39, cut into 16 cards, with instructions. Enough for one per learner.

TIME:
30 minutes for the activity, 10 minutes for class discussion.

NOTES

- This activity is suitable for use as a Main or Homework activity.
- Learners must be aware of these key words/concepts before attempting the task: antibiotics, immunisation, population, tuberculosis.

SUGGESTED ANSWERS

Studies of Egyptian mummies from 2400 BC showed signs of tuberculosis infection. TB was called 'consumption', it was considered an incurable disease.
In 1699, Italian doctors first wrote about tuberculosis being an infectious disease of the lung. However, most doctors continued to think that TB developed inside individuals.
Hermann Brehmer published a paper in 1854 stating that TB could be cured through rest, good diet and fresh air.
In 1882, Robert Koch discovered a staining technique that enabled him to see *Mycobacterium tuberculosis*.
In 1920s, French scientists Calmette and Guerin made a culture that later led to the BCG (Bacille-Calmette-Guerin) vaccination.
A team of scientists isolated an effective anti-TB antibiotic, actinomycin, in 1940. However, this proved to be too toxic for use in humans or animals.
In 1943, the antibiotic streptomycin was purified from the bacterium *Streptomyces griseus*. In 1944, the antibiotic was given to a critically ill TB patient, who then recovered successfully.
In 1953, vaccinations against TB were introduced, the BCG. Soon after, this jab was given routinely to school children in the UK.
In 1996, the World Health Organisation (WHO) estimated that 8 million people a year get TB, 95% of which are in developing countries. About 3 million people a year die from TB.
In 2005, the schools' BCG immunisation programme was stopped following continued decline in TB rates in the UK population. However, some 'high risk' groups are still given the jab.

EXTENSION SUGGESTION

Research and add own images and events.

HEALTH, MEDICINE AND DRUGS
TB TIMELINE

39

TASK

Read through the statements and arrange them in the correct chronological order on the timeline below.

Timeline		
0 AD	In 1996, the World Health Organisation (WHO) estimated that 8 million people a year get TB, 95% of which are in developing countries. About 3 million people a year die from TB.	
	In 1699, Italian doctors first wrote about tuberculosis being an infectious disease of the lung. However, most doctors continued to think that TB developed inside individuals.	
1600 AD	In 1920s, French scientists Calmette and Guerin made a culture that later led to the BCG (Bacille-Calmette-Guerin) vaccination.	
	Hermann Brehmer published a paper in 1854 stating that TB could be cured through rest, good diet and fresh air.	
1700 AD		
	In 1943, the antibiotic streptomycin was purified from the bacterium *Streptomyces griseus*. In 1944, the antibiotic was given to a critically ill TB patient, who then recovered successfully.	
	A team of scientists isolated an effective anti-TB antibiotic, actinomycin, in 1940. However, this proved to be too toxic for use in humans or animals.	
1800 AD		
	In 1882, Robert Koch discovered a staining technique that enabled him to see *Mycobacterium tuberculosis*.	
1900 AD	In 2005, the schools' BCG immunisation programme was stopped following continued decline in TB rates in the UK population. However, some 'high risk' groups are still given the jab.	
	In 1953, vaccinations against TB were introduced, the BCG. Soon after, this jab was given routinely to school children in the UK.	
2000 AD	Studies of Egyptian mummies from 2400 BC showed signs of tuberculosis infection. TB was called 'consumption,' it was considered an incurable disease.	

KS4 BIOLOGY HOW SCIENCE WORKS: COPYMASTER
Biology HSW © Badger Publishing Ltd

HEALTH, MEDICINE AND DRUGS
CHILDHOOD VACCINATIONS: SOCIAL, ECONOMIC AND ETHICAL ISSUES

40

KS4 NATIONAL CURRICULUM HSW LINK

4. *Applications and implications of science*
   b. to consider how and why decisions about science and technology are made, including those that raise ethical issues, and about the social, economic and environmental effects of such decisions

RESOURCES:
Task Sheet 40 cut into 6 cards, with instructions. Enough for one between two or four learners.

TIME:
10 minutes for the activity, 10 minutes for class discussion.

NOTES

- This activity is suitable for use as a Starter, Main activity or Plenary.
- Learners must be aware of these key words/concepts before attempting the task: vaccination, inoculation, childhood diseases.

SUGGESTED ANSWERS

1. Childhood diseases can be fatal or cause serious health problems, so I make sure that my children are vaccinated. Ranj, mother of two — (Eth)
2. At least 90% of the population needs to be vaccinated to make sure that the diseases are kept at a non-threatening level. Louis, NHS doctor — (S)
3. It is much cheaper to vaccinate people on the NHS than it is to treat the effects of the horrendous childhood diseases. Morris, Health Minister — (Ec)
4. I am not going to have my child injected with diseases. Vaccinations can be unsafe and some children have life-threatening reactions to them. David, father — (Eth)
5. The use of vaccinations against childhood diseases in the UK has reduced the incidence of these diseases to negligible levels. Jasmine, Professor of Medicine — (S)
6. A few of my patients are concerned about the potential side-effects of giving vaccinations to children. However, I always tell them that the chances of this are far lower than the risk and effects of getting the disease. Tunde, GP — (S/Eth)

EXTENSION SUGGESTION

Form an opinion with a reason.

KS4 BIOLOGY HOW SCIENCE WORKS: Teacher notes

HEALTH, MEDICINE AND DRUGS
CHILDHOOD VACCINATIONS: SOCIAL, ECONOMIC AND ETHICAL ISSUES

40

TASK

Discuss each statement card and decide whether it is a social, economic or ethical argument about childhood vaccinations.

- **Social argument** – concerned with the effects on everyone in society.
- **Economic argument** – concerned with costs and money.
- **Ethical argument** – concerned with what is considered right or wrong.

Childhood diseases can be fatal or cause serious health problems, so I make sure that my children are vaccinated. **Ranj, mother of two**	At least 90% of the population needs to be vaccinated to make sure that the diseases are kept at a non-threatening level. **Louis, NHS doctor**
It is much cheaper to vaccinate people on the NHS than it is to treat the effects of the horrendous childhood diseases. **Morris, Health Minister**	I am not going to have my child injected with diseases. Vaccinations can be unsafe and some children have life-threatening reactions to them. **David, father**
The use of vaccinations against childhood diseases in the UK has reduced the incidence of these diseases to negligible levels. **Jasmine, Professor of Medicine**	A few of my patients are concerned about the potential side-effects of giving vaccinations to children. However, I always tell them that the chances of this are far lower than the risk and effects of getting the disease. **Tunde, GP**

HEALTH, MEDICINE AND DRUGS
STEM CELL THERAPY: BENEFITS, DRAWBACKS AND RISKS

41

KS4 NATIONAL CURRICULUM HSW LINK

4. Applications and implications of science
   a. about the use of contemporary scientific and technological developments and their benefits, drawbacks and risks

RESOURCES:
Task Sheet 41 cut into 9 cards, with instructions. Enough for one between two or four learners.

TIME:
10 minutes for the activity, 10 minutes for class discussion.

NOTES

- This activity is suitable for use as a Starter, Main activity or Plenary.
- Learners must be aware of these key words/concepts before attempting the task: benefit, drawback, risk, stem cells.

SUGGESTED ANSWERS

Below is the intention of each statement: *Benefits: B, Drawbacks: D, Risks: R*

1. Stem cells are capable of turning into any type of cell. (B)
2. Some of the most useful and versatile stem cells are extracted from embryos. (D)
3. A human egg has the same moral status as a human, so using embryonic cells is the equivalent to murder. (D)
4. Stem cell therapy may increase the chance of developing cancer. (R)
5. Stem cells could be used to test the effects of new drugs. (B)
6. Stem cells are rare and hard to find in the adult body. (D)
7. Stem cells from adults can be used which do not harm the donor. (B)
8. Transplants using stem cell therapy may pass on viruses and other diseases to the recipient. (R)
9. Stem cells have the potential to cure a wide range of diseases. (B)

EXTENSION SUGGESTION

List any questions you have about the statements.

Health, Medicine and Drugs
Stem Cell Therapy: Benefits, Drawbacks and Risks

41

Task

Discuss each statement card and decide whether it is a benefit, drawback or a risk of stem cells to treat inherited conditions.

- A **benefit** is something that generally has a good effect on people.
- A **drawback** is something that is a hindrance or is the 'downside'.
- A **risk** is a possible danger or source of harm.

Stem cells are capable of turning into any type of cell.	Some of the most useful and versatile stem cells are extracted from embryos.	A human egg has the same moral status as an adult human, so using embryonic cells is the equivalent to murder.
Stem cell therapy may increase the chance of developing cancer.	Stem cells could be used to test the effects of new drugs.	Stem cells are rare and hard to find in the adult body.
Stem cells from adults can be used which do not harm the donor.	Transplants using stem cell therapy may pass on viruses and other diseases to the recipient.	Stem cells have the potential to cure a wide range of diseases.

ENVIRONMENT
SAMPLING: QUADRAT, TRANSECT OR SATELLITE IMAGING?

42

KS4 NATIONAL CURRICULUM HSW LINK

1. Data, evidence, theories and explanations
   a. how scientific data can be collected and analysed.

RESOURCES:
Task Sheet 42, enough for one between two learners.

TIME:
5-10 minutes for the activity, 5 minutes for the discussion.

NOTES

- This can be used as a Starter or Plenary, or a Homework activity.
- Learners must be aware of these key words/concepts before attempting the task: sampling, quadrat.

SUGGESTED ANSWERS

The purpose of this activity is to stimulate discussion about different approaches to sampling. Learners will hopefully recognise that some sampling can be carried out in more than one way.

1. Elephant damage to trees in an African woodland. (T)
2. Number of dandelions on the football field. (Q)
3. Effect of trampling in a Lake District National Park. (T/Q)
4. Extent of deforestation in a South American tropical rainforest. (S)
5. Number of rare medicinally important trees in a North American woodland. (T)
6. Number of weeds growing on an organic farm field. (Q)
7. Vegetation mapping of a Zambian National Park. (Q/T/S)
8. Effect of climate change on mangrove forests. (T/S)

EXTENSION SUGGESTION

Think of other uses of satellite imaging.

ENVIRONMENT
SAMPLING: QUADRAT, TRANSECT OR SATELLITE IMAGING?

42

Scientists and ecologists often need to study the plants growing in a particular environment. Three methods of sampling are: quadrats, line transects and satellite imaging.

Quadrat method
Quadrats can be used in two ways. In the first, a quadrat of a known area is placed randomly in the habitat and the number or types of plants counted. The quadrat is then randomly placed somewhere else in the habitat and the counting is continued until a set number of quadrat areas have been studied.

The second method involves dividing the area into squares on a grid. Squares on the grid are randomly chosen. The plants in the selected squares are counted.

Line or belt transect method
A transect line is made using a rope or tape marked and numbered at 0.5m or 1m intervals, all the way along its length. The rope is laid across the area to be studied. The species touching the line (line transect) or within a certain distance from the line (belt transect) are recorded along the whole length of the line.

Satellite imaging
Orbiting satellites take digital images of the Earth's surface. The images can then be analysed to identify different vegetation types and how they are changing.

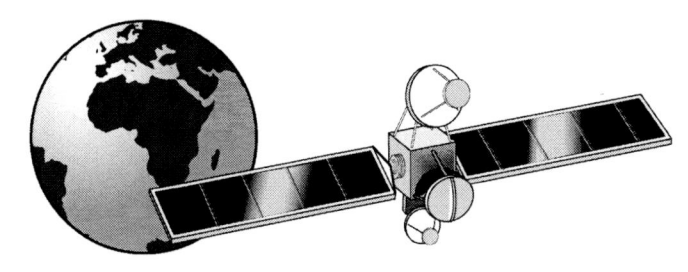

TASK

Discuss Which of these methods could be used to investigate the following:

1. Elephant damage to trees in an African woodland.
2. Number of dandelions on the football field.
3. Effect of trampling in a Lake District National Park.
4. Extent of deforestation in a South American tropical rainforest.
5. Number of rare medicinally important trees in a North American woodland.
6. Number of weeds growing on an organic farm field.
7. Vegetation mapping of a Zambian National Park.
8. Effect of climate change on mangrove forests.

ENVIRONMENT
ANALYSING GLOBAL WARMING

43

KS4 NATIONAL CURRICULUM HSW LINK

1. Data, evidence, theories and explanations
   c. how explanations of many phenomena can be developed using scientific theories, models and ideas

RESOURCES:
Task Sheet 43, enough for one between two learners.

TIME:
10 minutes for the activity, 10 minutes for the discussion.

NOTES

- This can be used as an extended Starter, Main or Homework activity, or a Plenary.
- Learners must be aware of these key words/concepts before attempting the task: global warming, the greenhouse effect.

SUGGESTED ANSWERS

A. What is the trend in carbon dioxide concentrations over the past 500 years?
   More or less level for 400 years, then a big increase.

B. What do you think has caused these changes in carbon dioxide concentrations?
   Increased burning of fossil fuels, increased agriculture, etc.

C. What is the trend in global temperature over the past 500 years?
   More or less level for 400 years, then a big increase.

D. Most scientists now agree that the increase in global temperatures is due to increasing concentrations of carbon dioxide in the atmosphere. Describe and explain how carbon dioxide concentrations affect global temperature.
   Draw out discussion of correlation and cause and effect as well as the concept of more thermal energy being absorbed by the atmosphere.

EXTENSION SUGGESTION

Can learners think of other explanations for the increase in global temperatures?

ENVIRONMENT
ANALYSING GLOBAL WARMING

43

TASK

Look at the two graphs. Figure 1 one shows changes in carbon dioxide concentration in the atmosphere over the past 500 years. Figure 2 shows global temperature changes over the same time period.

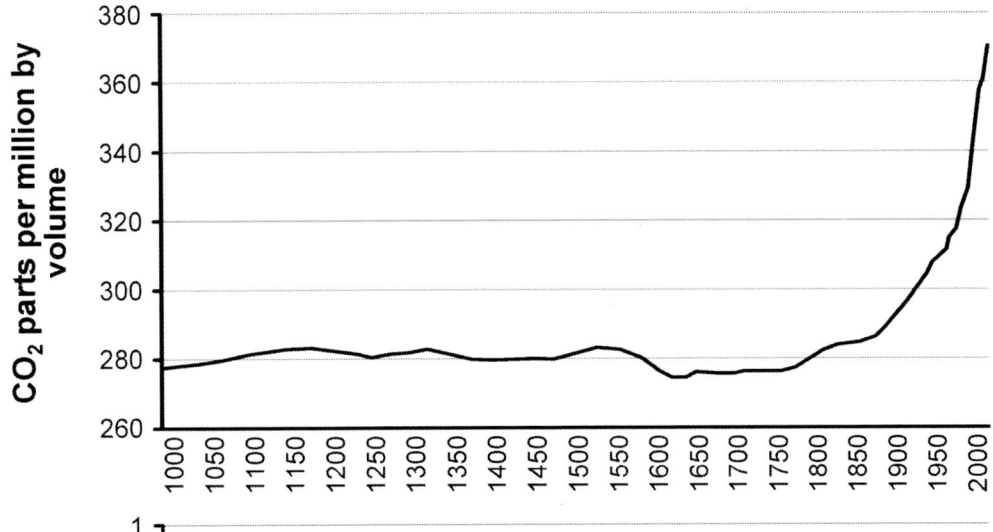

Figure 1

Figure 2

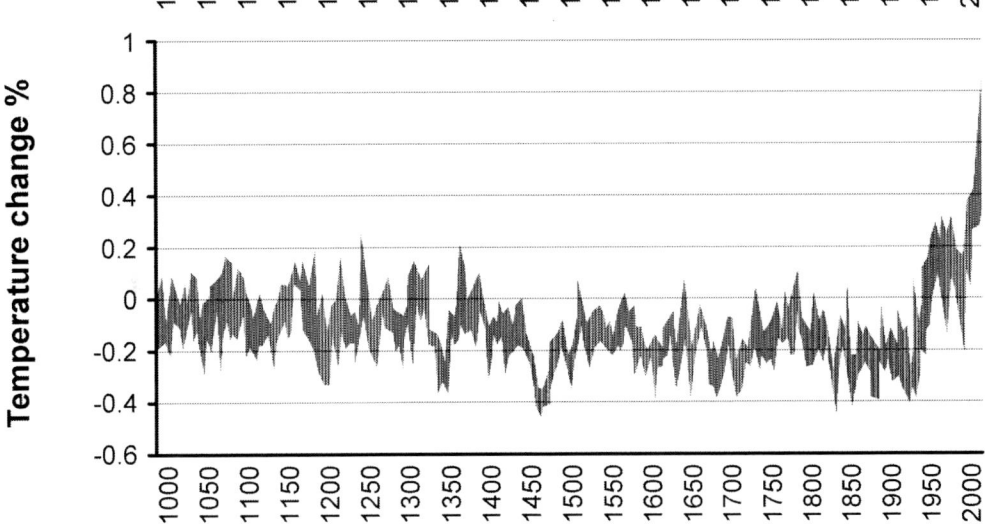

Discuss

A. What is the trend in carbon dioxide concentrations over the past 500 years?
B. What do you think has caused these changes in carbon dioxide concentrations?
C. What is the trend in global temperature over the past 500 years?
D. Most scientists now agree that the increase in global temperatures is due to increasing concentrations of carbon dioxide in the atmosphere. Describe and explain how carbon dioxide concentrations affect global temperature.

KS4 BIOLOGY HOW SCIENCE WORKS: COPYMASTER

Biology HSW © Badger Publishing Ltd

ENVIRONMENT
QUADRAT DATA COMPARISON

44

KS4 NATIONAL CURRICULUM HSW LINK

1. Data, evidence, theories and explanations
   a. how scientific data can be collected and analysed

RESOURCES:
Task Sheet 44, enough for one between two learners.

TIME:
10 minutes for the activity, 5 minutes for the discussion.

NOTES

- This can be used as an extended Starter, Main or Homework activity, or Plenary.
- Learners must be aware of these key words/concepts before attempting the task: the sampling techniques involved, quadrat, plantain, herbicide.

SUGGESTED ANSWERS

A. Calculate the average number of plants per quadrat for each plant species.

Group A:

Plant species	Average/quadrat
Dandelion	1.1
Groundsel	1.0
Plantain	3.5

Group B:

Plant species	Average/quadrat
Dandelion	0.5
Groundsel	0.5
Plantain	0.7

B. The quadrat measures 50cm x 50cm. The field is 50m long by 80m wide.
   i. How many m^2 is the field?   ii. How many quadrats will fit onto the field?
   i. $4,000m^2$
   ii. 16,000 quadrats

C. Calculate how many plantains there are on the field using the data from:
   i. Group A
   ii. Group B
   i. 56,000 plantains
   ii. 11,200 plantains

D. Can you explain why the groups arrived at such different answers for the population size of plantains on the field?
   The students using the random sampling method selected areas of field which were heavily populated or they sampled only a small area of the field.

E. Which method of sampling, 1 or 2, gave the most reliable results?
   Method 2 is the most reliable method because the squares were chosen randomly before the students reached the field so they were not biased by the plants in the field. This gives the results greater validity.

EXTENSION SUGGESTION

List the pros and cons of each sampling method.

KS4 BIOLOGY HOW SCIENCE WORKS: TEACHER NOTES

ENVIRONMENT
QUADRAT DATA COMPARISON

44

A school caretaker is concerned about weeds growing on the school cricket pitch. He knows he will need to use a weed killer to destroy the weeds but he is not sure about the types and numbers of weeds growing. A class of students decide to help him by counting the number and types of weeds on the field. They will use two sampling techniques they have learnt in science.

Method 1: Random sampling
A quadrat of a known area is placed randomly on the field. The types of weeds and their numbers are recorded. The quadrat is then randomly placed on another part of the field and the weeds counted again. The data is recorded in a table of results. This procedure is repeated a number of times.

Method 2: Grid method
The field is divided into a grid, each square is the size of the quadrat. The squares are numbered. Square numbers are randomly selected. The corresponding area on the field is then located and the types and numbers of weeds are counted. This procedure is carried out with each square number selected.

The students' data will enable the caretaker to make informed decisions about the type of herbicide to use, and the concentration and number of applications needed.

Group A

Plant species	\multicolumn{10}{c}{Quadrat}	Average/ quadrat
	1	2	3	4	5	6	7	8	9	10	
Dandelion	0	2	1	2	1	1	1	1	0	2	
Groundsel	1	0	0	1	2	3	3	0	0	0	
Plantain	4	4	5	4	3	4	3	2	3	3	

Group B

Plant species	1	2	3	4	5	6	7	8	9	10	Average/ quadrat
Dandelion	0	0	1	0	1	0	0	1	0	2	
Groundsel	1	0	0	1	1	2	0	0	0	0	
Plantain	2	1	0	0	0	0	0	2	1	1	

TASK

A. Calculate the average number of plants per quadrat for each plant species.
B. The quadrat measures 50cm x 50cm. The field is 50m long by 80m wide.
   i. How many m^2 is the field?)
   ii. How many quadrats will fit onto the field?
   (tip: area of field ÷ area of one quadrat
C. Calculate how many plantains there are on the field using the data from:
   i. Group A   ii. Group B
D. Can you explain why the groups arrived at such different answers for the population size of plantains on the field?
E. Which method of sampling, 1 or 2, gave the most reliable results?

ENVIRONMENT
LICHENS AND AIR POLLUTION

45

KS4 NATIONAL CURRICULUM HSW LINK

3. Communication skills
   a. recall, analyse, interpret, apply and question scientific information or ideas
   b. use both qualitative and quantitative approaches
   c. present information, develop an argument and draw a conclusion, using scientific, technical and mathematical language, conventions and symbols, and ICT tools

RESOURCES:
Task Sheet 45, enough for one between two learners.

TIME:
10 minutes for the activity, 10 minutes for the discussion.

NOTES

- This can be used as a Starter or Plenary to stimulate discussion on theories of the use of lichens as bio-indicators.
- Learners must be aware of these key words/concepts before attempting the task: lichen, bio-indicator, sulphur dioxide, alga, cyanobacterium.

SUGGESTED ANSWERS

A. What does the map tell you about the geographical distribution of lichen in the UK?
   There is an inverse relationship between human population and lichen population. The black areas on the map show no lichen present and are also areas of high human population. So, where there is a lot of human activity, there are no lichens present. Pollution sensitive lichens are found where the human population is sparse.

B. How does the distribution of lichen relate to the human population of the UK? What does this tell you about sources of air pollution in the UK?
   There is more air pollution where people live.

C. What sort of lichens would you expect to find where you live?
   Find locality on map.

D. What is the source of sulphur dioxide pollution?
   The main source is the burning of fossil fuels in power stations, but other sources include smelting, manufacture of sulphuric acid, conversion of wood pulp to paper and incineration of refuse.

EXTENSION SUGGESTION

Suggest ways to reduce air pollution.

ENVIRONMENT
LICHENS AND AIR POLLUTION

45

A lichen is not a single organism. In fact, a lichen is a symbiotic association between a fungus and a photosynthetic partner (an alga or cyanobacterium). The fungus gains food from its photosynthetic partner and provides its partner with water, minerals and shelter. Lichens are usually seen as crusts or hairy, shrubby or leaf-like structures growing on bare surfaces such as rocks, soil and tree bark.

Lichens are useful to humans because many are sensitive to sulphur dioxide in the atmosphere – lichens can be used as air pollution bio-indicators. Generally, less lichens present in a habitat means higher levels of sulphur dioxide pollution in the air. But the type of lichen changes as air pollution increases (see table below).

Zone	Type of lichen present	Levels of air pollution	Typical lichen
1	No lichen present	Very high	
2	Crust-forming	Moderate	Crusts on rock *Lecanora*
3	Leaf-like	Low	Leafy lichen *Parmelia*
4	Hairy and filamentous	Very clean air	Hairy lichen *Usnea*

TASK

The map shows the distribution of lichen in the UK.

A. What does the map tell you about the geographical distribution of lichen in the UK?
B. How does the distribution of lichen relate to the human population of the UK? What does this tell you about sources of air pollution in the UK?
C. What sort of lichens would you expect to find where you live?
D. What is the source of sulphur dioxide pollution?

The black areas on the map show where lichens are not found (Zone 1).

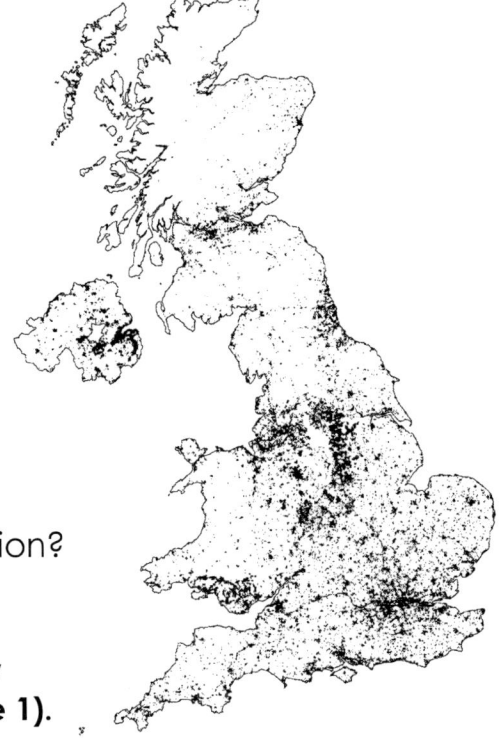

ENVIRONMENT
FERTILISER CHOICES

46

KS4 NATIONAL CURRICULUM HSW LINK

1. Data, evidence, theories and explanations
   a. how scientific data can be collected and analysed
   b. how interpretation of data, using creative thought, provides evidence to test ideas and develop theories
   c. how explanations of many phenomena can be developed using scientific theories, models and ideas

RESOURCES:
Task Sheet 46, enough for one set between 2-4 learners.

TIME:
10 minutes for the activity, 10 minutes for the discussion.

NOTES

- This task is suitable as a Starter or Plenary, or Main or Homework activity.
- Learners must be aware of these key words/concepts before attempting the task: fertiliser, nutrients.

SUGGESTED ANSWERS

	Fertiliser A	Fertiliser B	Fertiliser C
N:P:K (%)	25:15:15	20:20:20	11:9:30
	Lawn feed: high nitrogen for leaf and stem growth.	House plant feed: has equal amounts of N:P:K to encourage all over growth.	Tomato feed: high potassium for fruit growth.

EXTENSION SUGGESTION

The gardener finds some pelleted organic chicken manure with N:P:K values of 6:5:5. He thinks it might be better for the environment to use the pelleted manure on his lawn and tomatoes. Would the pelleted manure promote healthy plant growth on his lawn and tomatoes? Can you explain your answer?

[Chicken manure pellets will not promote growth as effectively as the fertiliser above due to the lack of high levels of nitrogen and potassium.]

KS4 BIOLOGY HOW SCIENCE WORKS: TEACHER NOTES

ENVIRONMENT
FERTILISER CHOICES

46

A gardener buys different fertilisers for different purposes. He buys one to feed his lawn, one to feed his tomatoes and one to feed his house plants. He stores the packets of fertiliser in his shed. However, when he comes to use the fertilisers, he discovers a mouse has eaten the labels of the boxes of fertiliser. All that remains of the information labels are the percentage ratios of nitrogen: phosporus: potassium (N:P:K) in the fertiliser formulations.

	Fertiliser A	Fertiliser B	Fertiliser C
N:P:K (%)	25:15:15	20:20:20	11:9:30

The gardener finds this information in a gardening book.

Nutrients	Plants need these nutrients for:
Nitrogen (N)	Essential for luxuriant leaf and stem growth; encourages plants to have a deep green colour; increases protein content of edible plants.
Phosphorus (P)	Essential for healthy root growth; required for flower and fruit production; encourages seed growth.
Potassium (K)	Essential for healthy flowers and fruits; encourages vigorous growth; required for good disease resistance.

TASK

Using the information above, help the gardener decide which fertiliser should be used to:

A. Feed his lawn.
B. Feed his tomatoes for lots of flavoursome fruit.
C. Feed his house plants for healthy plants with good all round growth.

ENVIRONMENT
MALARIA AND DDT

47

KS4 NATIONAL CURRICULUM HSW LINK

3. Communication skills
   a. recall, analyse, interpret, apply and question scientific information or ideas
   b. use both qualitative and quantitative approaches
   c. present information, develop an argument and draw a conclusion, using scientific, technical and mathematical language, conventions and symbols, and ICT tools

RESOURCES:
Task Sheet 47, enough for one between two learners.

TIME:
10 minutes for the activity, 10 minutes for the discussion.

NOTES

- This can be used as a Starter or Plenary to stimulate discussion on theories about the use of pesticides.
- Learners must be aware of these key words/concepts before attempting the task: pesticide, resistance, interpreting graphs.

SUGGESTED ANSWERS

A. What is the relationship between cases of malaria and households sprayed?
   Spraying reduces the number of cases of malaria.

B. Suggest why the number of houses sprayed decreased rapidly between 1959 and 1962.
   Because people became aware of the harmful effects of DDT coupled with low numbers of cases of malaria.

C. How many households should have been sprayed with DDT to keep cases of malaria at approximately 20,000 cases per year i. before 1988 ii. after 1988? Make sure you can explain your answers.
   i. 2.5–3 million   ii. 1.5-2 million

D. Can you think of any reasons other than the use of DDT which caused cases of malaria to decrease rapidly after 1988?
   Higher standards of living conditions, better housing, sleeping under mosquito nets and reduction in amount of standing water.

EXTENSION SUGGESTION

Many insects have become resistant to DDT. Predict how DDT resistance would have affected the number of cases of malaria.

ENVIRONMENT
MALARIA AND DDT

47

Malaria is a disease of human blood caused by single celled organisms called trypanosomes. Trypanosomes are transmitted from one person to another by mosquitoes. Mosquitoes are small biting insects that live and breed by still pools of water.

Malaria is a very serious disease that can result in death. It is common in developing countries.

Malaria can be controlled using insecticides to reduce mosquito populations. Perhaps the most famous insecticide used against mosquitoes is DDT (dichloro-diphenyl-trichloroethane). The insecticidal properties of DDT were discovered in 1939 and, within a few years, DDT was being used to control a number of insects, including mosquitoes and fleas, which carry typhus. However, despite its efficacy against mosquitoes, DDT has been banned in many countries because of its harmful side effects on the environment.

Figure 1. Number of cases of malaria and the number of houses sprayed with DDT

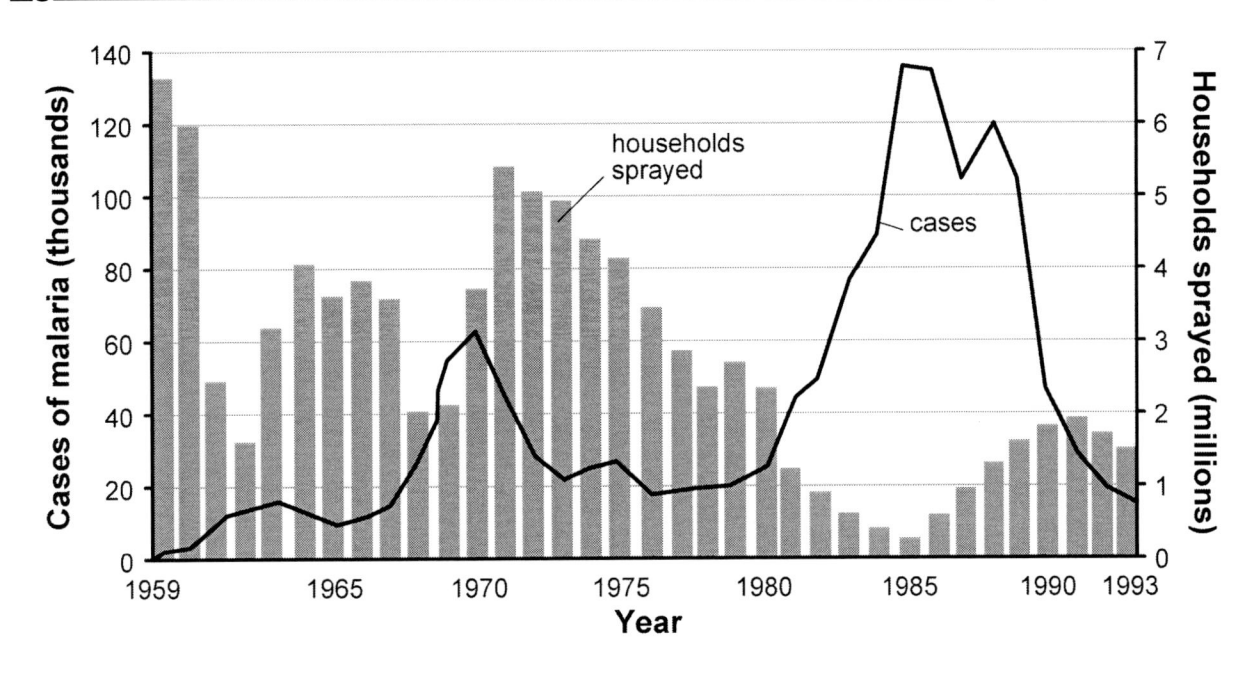

TASK

A. What is the relationship between cases of malaria and households sprayed?
B. Suggest why the number of houses sprayed decreased rapidly between 1959 and 1962.
C. How many households should have been sprayed with DDT to keep cases of malaria at approximately 20,000 cases per year i. before 1988 ii. after 1988? Make sure you can explain your answers.
D. Can you think of any reasons other than the use of DDT which caused cases of malaria to decrease rapidly after 1988?

Environment
DDT: How Ideas Changed

48

KS4 National Curriculum HSW link

4. Applications and implications of science
   c. how uncertainties in scientific knowledge and scientific ideas change over time and about the role of the scientific community in validating these changes

Resources:
Task Sheet 48, cut into 13 cards, with instructions. Enough for one per learner.

Time:
10 minutes for the activity, 10 minutes for class discussion.

Notes

- This can be used as a stimulus activity when discussing the impact of humans on the environment. This activity could also be used as a Plenary or Homework activity.
- Learners must be aware of these key words before attempting the task: pesticide.

Suggested answers

DDT (dichloro-diphenyl-trichloroethane) was synthesised for the first time by German Othmar Zeidler in 1874.
In 1939, Swiss scientist Paul Mueller discovered that DDT was a very effective insecticide. Nine years later, in 1948, he was awarded the Nobel Prize for Physiology and Medicine.
Between 1943 and 1945, DDT was used in the South Pacific to eliminate insect pests, including fleas carrying typhus. Its insecticidal properties benefited both the military and civilians.
For the next four years, until 1949, DDT was used extensively to control mosquitoes across Europe and North America, preventing outbreaks of malaria and eliminating malaria from countries such as Italy.
In 1954, the World Health Organisation (WHO) implemented a worldwide malaria eradication programme by spraying the interior of dwellings with DDT.
During 1956, fifty different species of insect were reported to be resistant to DDT in the US. This coincided with reports of DDT causing egg thinning in birds of prey, almost resulting in the extinction of the bald eagle.
In 1962, *Silent Spring* by Rachel Carson was published, in which she urged a more responsible use of pesticides. The publication of *Silent Spring* coincided with peak production of DDT in the US (averaging 200 million tonnes per year).
1969 saw residues of DDT found around the world. In 1972, the WHO reported that malaria had been eradicated from 37 countries. The use of DDT was banned in many parts of the world (although its use still continues to the present).

Extension suggestion

Explain why some countries still use DDT.

ENVIRONMENT
DDT: HOW IDEAS CHANGED

48

TASK

Read through the statements on the cards. Place the cards in the correct order on the timeline. Note how attitudes towards DDT changed over a relatively short period of time.

Timeline

- 1800 AD
- 1900 AD
- 2000 AD

1969 saw residues of DDT found around the world. In 1972, the WHO reported that malaria had been eradicated from 37 countries. The use of DDT was banned in many parts of the world (although its use still continues to the present).	
In 1939, Swiss scientist Paul Mueller discovered that DDT was a very effective insecticide. Nine years later, in 1948, he was awarded the Nobel Prize for Physiology and Medicine.	
Between 1943 and 1945, DDT was used in the South Pacific to eliminate insect pests, including fleas carrying typhus. Its insecticidal properties benefited both the military and civilians.	
For the next four years, until 1949, DDT was used extensively to control mosquitoes across Europe and North America, preventing outbreaks of malaria and eliminating malaria from countries such as Italy.	
During 1956, fifty different species of insect were reported to be resistant to DDT in the US. This coincided with reports of DDT causing egg thinning in birds of prey, almost resulting in the extinction of the bald eagle.	
In 1954, the World Health Organisation (WHO) implemented a worldwide malaria eradication programme by spraying the interior of dwellings with DDT.	
In 1962, *Silent Spring* by Rachel Carson was published, in which she urged a more responsible use of pesticides. The publication of *Silent Spring* coincided with peak production of DDT in the US (averaging 200 million tonnes per year).	
DDT (dichloro-diphenyl-trichloroethane) was synthesised for the first time by German Othmar Zeidler in 1874.	

KS4 BIOLOGY HOW SCIENCE WORKS: COPYMASTER

Biology HSW © Badger Publishing Ltd

ENVIRONMENT
GLOBAL WARMING: FACT OR OPINION?

49

KS4 NATIONAL CURRICULUM HSW LINK

4. Applications and implications of science
   c. how uncertainties in scientific knowledge and scientific ideas change over time and about the role of the scientific community in validating these changes

RESOURCES:
Task Sheet 49 cut into 9 cards, with instructions. Enough for one set between 2-4 learners.

TIME:
10 minutes for the activity, 10 minutes for class discussion.

NOTES

- This activity is suitable as a Starter, Main activity or Plenary.
- You may need to define the terms: global warming, greenhouse effect, solar activity.

SUGGESTED ANSWERS

1. Global warming is a result of human activity like burning fossil fuels. *(Opinion)*
2. The ice caps are melting due to global warming. *(Fact)*
3. Global warming is caused by increasing levels of carbon dioxide in the atmosphere. *(Opinion)*
4. Global warming is due to increased solar activity. *(Opinion)*
5. Global warming will result in changes in weather patterns. *(Fact)*
6. Summers in the UK are becoming wetter. *(Fact)*
7. Temperatures have remained relatively constant over the past 10 years or so. *(Fact)*
8. Global warming is due to the Earth's natural cycle of warming and cooling. *(Opinion)*
9. The average temperature of the Earth has increased over the past century. *(Fact)*

EXTENSION SUGGESTION

List any questions you have about the statements.

KS4 BIOLOGY HOW SCIENCE WORKS: Teacher notes

ENVIRONMENT
GLOBAL WARMING: FACT OR OPINION?

49

TASK

Discuss each statement card and decide whether it is a fact or an opinion about global warming. Place the cards into two piles: facts about global warming and opinions about global warming.

- A **fact** is something that has been proved true.
- An **opinion** is something that someone feels or believes, whether or not it is true.

Global warming is a result of human activity like burning fossil fuels.	The ice caps are melting due to global warming.	Global warming is caused by increasing levels of carbon dioxide in the atmosphere.
Global warming is due to increased solar activity.	Global warming will result in changes in weather patterns.	Summers in the UK are becoming wetter.
Temperatures have remained relatively constant over the past 10 years or so.	Global warming is due to the Earth's natural cycle of warming and cooling.	The average temperature of the Earth has increased over the past century.

ENVIRONMENT
PESTICIDES: BENEFITS, DRAWBACKS AND RISKS

50

KS4 NATIONAL CURRICULUM HSW LINK

4. Applications and implications of science
   a. about the use of contemporary scientific and technological developments and their benefits, drawbacks and risks

RESOURCES:
Task Sheet 50 cut into 8 cards, with instructions. Enough for one between two or four learners.

TIME:
10 minutes for the activity, 10 minutes for class discussion.

NOTES

- This activity is suitable for use as a Starter, Main activity or Plenary.
- Learners must be aware of these key words/concepts before attempting the task: insecticides, genetic resistance, bio-accumulate, benefit, drawback and risk.

SUGGESTED ANSWERS

Below is the intention of each statement: *Benefits: B, Drawbacks: D, Risks: R.*

1. Pesticides can kill a large number of harmful insects very quickly. (B)
2. Pests can become resistant to pesticides after prolonged use. (D)
3. Many chemical pesticides can be dangerous to humans if they are breathed in or absorbed through the skin. Worldwide between 1 and 5 million people have died from pesticide poisoning (about 220,000 per year). (D/R)
4. Pesticides can be used to kill disease-causing pests such as mosquitoes that cause malaria, saving millions of human lives. (B)
5. Pesticides can be used to kill insects that can destroy crops, increasing the yield by up to 40%. (B)
6. Some pesticides can bio-accumulate (build up) in food chains, causing reproductive problems in tertiary consumers. (D/R)
7. Many pesticides are not specific and kill all insects in a field, not just the pest species. This reduces food for birds. (D)
8. With pesticides that are applied to fields by spraying from a small plane, only 10% reaches the crop and only 1-5% reaches the target pest. (D)

EXTENSION SUGGESTION

List some alternatives to the use of pesticides.

ENVIRONMENT
PESTICIDES: BENEFITS, DRAWBACKS AND RISKS

50

TASK

Discuss each statement card and decide whether it is a benefit, drawback or a risk of using pesticides.

- A **benefit** is something that generally has a good effect on people.
- A **drawback** is something that is a hindrance or is the 'downside'.
- A **risk** is a possible danger or source of harm.

Pesticides can kill a large number of harmful insects very quickly.	Pests can become resistant to pesticides after prolonged use.
Many chemical pesticides can be dangerous to humans if they are breathed in or absorbed through the skin. Worldwide between 1 and 5 million people die from pesticide poisoning.	Pesticides can be used to kill disease-causing pests such as mosquitoes that cause malaria, saving millions of human lives.
Pesticides can be used to kill insects that can destroy crops, increasing the yield by up to 40%.	Some pesticides can bio-accumulate (build up) in food chains, causing reproductive problems in tertiary consumers.
Many pesticides are not specific and kill all insects in a field, not just the pest species. This reduces food for birds.	With pesticides that are applied to fields by spraying from a small plane, only 10% reaches the crop and only 1-5% reaches the target pest.

KS4 BIOLOGY HOW SCIENCE WORKS: COPYMASTER

Biology HSW © Badger Publishing Ltd

Badger Publishing Limited
15 Wedgwood Gate
Pin Green Industrial Estate
Stevenage, Hertfordshire SG1 4SU
Telephone: 01438 356907
Fax: 01438 747015
www.badger-publishing.co.uk
enquiries@badger-publishing.co.uk

Badger GCSE Science
How Science Works
Biology

First published 2008
ISBN 978 1 84691 307 5

Text © Andrew Grevatt and Dr. Deborah Shah-Smith 2007
Complete work © Badger Publishing Limited 2007

The right of Andrew Grevatt and Dr. Deborah Shah-Smith to be identified as author of this Work has been asserted by them in accordance with the Copyright, Designs and Patents Act 1988.

Once purchased, you may copy this book freely for use in your school.
The pages in this book are copyright, but copies may be made without fees or prior permission provided that these copies are used only by the institution which purchased the book. For copying in any other circumstances, prior written consent must be obtained from the publisher.

Publisher: David Jamieson
Editor: Paul Martin
Designer: Adam Wilmott
Illustrator: John Dillow (Beehive Illustration), Adam Wilmott

Cover photo: E. coli: Culture dish with Escherichia coli (E. coli) © WILL & DENI MCINTYRE/SCIENCE PHOTO LIBRARY.

Printed in the UK

BADGER SCIENCE
NEW FOR 2008
By Andrew Grevatt

Badger *Level-Assessed Tasks* Concepts

Y7 Concepts Teacher Book and CD	ISBN 978-1-84691-301-3
Y8 Concepts Teacher Book and CD	ISBN 978-1-84691-302-0
Y9 Concepts Teacher Book and CD	ISBN 978-1-84691-303-7

Badger *Level-Assessed Tasks* HSW

Y7 HSW Teacher Book and CD	ISBN 978-1-84691-304-4
Y8 HSW Teacher Book and CD	ISBN 978-1-84691-305-1
Y9 HSW Teacher Book and CD	ISBN 978-1-84691-306-8

How Science Works for KS4

Biology Teacher Book and CD	ISBN 978-1-84691-307-5
Chemistry Teacher Book and CD	ISBN 978-1-84691-309-9
Physics Teacher Book and CD	ISBN 978-1-84691-308-2

Also available from Badger Publishing

Badger *Assessment for Learning Tasks*
Andrew Grevatt and Anne Penfold

GCSE Core Topics Teacher Book and CD	ISBN 978-1-84424-952-7
Additional Topics Teacher Book and CD	ISBN 978-1-84691-207-8

Badger *Science Starters*
John Parker

Y7 Teacher Book	ISBN 978-1-85880-353-1
Y7 PDF CD	ISBN 978-1-84424-130-9
Y8 Teacher Book	ISBN 978-1-85880-354-8
Y8 PDF CD	ISBN 978-1-84424-131-6
Y9 Teacher Book	ISBN 978-1-85880-355-5
Y9 PDF CD	ISBN 978-1-84424-132-3

For the Interactive Whiteboard Mary Mather

Y7 SMART™	ISBN 978-1-84424-779-0
Y7 Promethean	ISBN 978-1-84424-780-6

See our full colour catalogue (available on request)
or visit our website for more information:
www.badger-publishing.co.uk.
Or visit our showroom and bookshop.

For details of the full range of books and resources from

Badger Publishing

including
- **Book Boxes** for 11-16 and Special Needs and Class sets of novels
- **KS3 Guided Reading** – Teacher Files and book packs
- **Between the Lines** – course exploring text types at KS3
- **Under the Skin** – progressive plays for KS3
- **Full Flight, Dark Flight, First Flight & Rex Jones** for reluctant readers
- **Brainwaves** – non-fiction to get your brain buzzing
- **SAT Attack** – Badger English and Maths Test Revision Guides
- **Badger KS3 Starters** for *Literacy, Maths* and *Science* and for the *Foundation Subjects* – History, Geography, Religious Education, Music, Design & Technology, Modern Foreign Languages
- **Main Activity: Problem Solved!** KS3 Maths problem solving course
- **Concepts** and **How Science Works** Level-Assessed Tasks for KS3 Science
- **Badger ICT** – lesson plans for KS3
- **Badger Music** – lesson plans for KS3
- **Building Blocks History** – complete unit for KS3
- **Thinking Together in Geography** – developing thinking skills
- **Multiple Learning Activities** – providing for different learning preferences
- **Beliefs and Issues** – Badger KS3 Religious Education course
- **Black** and **Asian Pride** and **British Role Model** poster sets
- **Dual Language** readers and **Full Flight Runway** for EAL

KEY STAGE 4
- **Badger KS4 Starters** for *Maths*
- **Science Assessment for Learning Tasks** for KS4
- **How Science Works** tasks for KS4
- **Badger GCSE Religious Studies** – illustrated text books
- **Surviving Citizenship @ KS4** – teaching guide

INTERACTIVE WHITEBOARD
- **Badger IAW Activities** for KS3 Music
- **Y7 Science Starters** for Smart™ Board and Promethean
- **Full Flight Guided Writing CD** – writing activities in Word
- **PDF CD** versions of many titles also now available.

See our full colour catalogue, visit our website or our showroom for more:
www.badger-publishing.co.uk

Badger Publishing Limited
15 Wedgwood Gate, Pin Green Industrial Estate,
Stevenage, Hertfordshire SG1 4SU
Telephone: 01438 356907   Fax: 01438 747015
enquiries@badger-publishing.co.uk